Farm Women Empowerment through Dairy Co-operative Societies

AF640330

NIPA® GENX ELECTRONIC RESOURCES & SOLUTIONS P. LTD.
New Delhi-110 034

www.nipaers.com
Browse, Search, Read & Buy...
Print Books
eChapters
eBooks
Publishers
Forthcoming
Subject Catalogues
New Titles
Online Resources on:
✓ Current Affairs
✓ Reasoning, Logic & Aptitude
✓ Competitive Examinations
✓ English Language Lab
✓ Writing & Pronunciation Tools
✓ Personality Development
✓ Online Programmes for Professionals
✓ Interview Preparation
Pay Using
PhonePe
Google Pay
paytm
PayPal
UPI
CC Avenue

Farm Women Empowerment through Dairy Co-operative Societies

Pitambar Swain
B.V.Sc.& A.H., M.V.Sc., Ph.D., M.B.A (HRM)
Professor, Department of Animal Husbandry and Extension
College of Veterinary Sciences and Animal Husbandry
Orissa University of Agriculture & Technology
Bhubaneswar - 751 003, Odisha

NIPA® GENX ELECTRONIC RESOURCES & SOLUTIONS P. LTD.
New Delhi-110 034

NIPA® GENX ELECTRONIC RESOURCES & SOLUTIONS P. LTD.

101,103, Vikas Surya Plaza, CU Block
L.S.C. Market, Pitam Pura, New Delhi-110 034
Ph : +91 11 27341616, 27341717, 27341718
E-mail: newindiapublishingagency@gmail.com
www: www.nipabooks.com

For customer assistance, please contact
Phone: + 91-11-27 34 17 17
Fax: + 91-11- 27 34 16 16
E-Mail: feedbacks@nipabooks.com

ISBN: 978-81-19103-92-8

Composed and Designed by NIPA®.

Preface

I am immensely happy to write this book on "Farm Woman Empowerment through Dairy Cooperative Societies". I got the inspiration to write this book from Prof. Dr.Chakradhar Sathapathy, the former Prof and Head, Dept.of Agriculture Extension and former Dean, Directorate of Extension Education, OUAT, Bhubaneswar. I am indebted to him for the milk of kindness he has shown in guiding me for my Ph.D programme. I owe my gratefulness to our Hon'ble Vice-chancellor, OUAT, Prof.D.P.Roy, for his illustrious inspiration to me and to Prof. P.K. Dehuri, Dean, C.V.Sc &A.H. for his constant moral support for writing this Book.

Many times, we are forgetting to understand the same old proverb "United we stand and Divided we fall" or to make it simpler "Unity is the Strength". People have enormous power, that too women. Cooperation is the base for sustenance of life. If this principle could be applied in our every walk of life, the sky would be the limit of the fruits of success. The women should be awaken for their unbelievable potential they have. This potential should be converted into the capability through appropriate intervention. The capable women folk are a sizable stock of human resource of our country. When appropriate opportunity along with need based support is provided that will encourage them to be more practical and productive in their respective field .This paves the way towards empowerment. When women are empowered physically, financially, spiritually, mentally, and above all socially, there would be tremendous change in the society thus leads to the development of the country.

Dairy Cooperative Society is one of the best alternative means for farm women to engage and empower themselves in all aspects of their life. Cow keeping and curd making are the traditional practice of rural women. Only this practice is to be channelized in a systematic cooperative mechanism for greater interest and better performance.

This is an attempt to bring into lime light that how farm women can be self- sustained socio- economically by adopting our days old tradition of cow keeping and curd making. In this Book steps have been taken to explain different aspects of functioning of Women Dairy Cooperatives Societies such as their performance, functioning, organization, entrepreneurial behaviour of members & constraints in functioning etc. with the support of research evidence. This book will be a great help to all those esteemed personalities working for Rural Development, Women Empowerment, Policy makers and more so all the NGOS working in Rural Development sectors.

I sincerely thank to all the authors, writers of different articles which I have referred in the course of writing this book.

I thank Sri S.K. Pattanaik, Artist, College of Veterinary Science and Animal Husbandry for taking pain for designing & organizing the book. Truley I will be morally elevated if this book will help my esteemed readesrs to satisfy their area of interest pertaining to the women dairy cooperative societies.

Author

Contents

Abbreviation

A.I.	:	Artificial Insemination
ACWA	:	Agricultural Cooperatives Women Association
BQ	:	Black Quarter
CDP	:	Community Development Programme
DPAP	:	Drought Prone Area Programme
DRDA	:	District Rural Development Agency
DWCRA	:	Development of Women and Children in Rural Areas
EIS	:	Economic Intelligence Service
FMD	:	Foot and Mouth Disease
GOI	:	Government of India
HS	:	Hemorrhagic Septicemia
ICDP	:	Intensive Cattle Development Project
ICDS	:	Intensive Child Development Scheme
ICMR	:	Indian Council of Medical Research
IDF	:	Indian Dairy Federation
ILDP	:	Integrated Livestock Development Project
IRDP	:	Integrated Rural Development Programme
ISPO	:	Indo Swiss Project
KBK	:	Kalahandi, Bolangir and Koraput
KPA	:	Key Performance Areas
LAC	:	Livestock Aid Centre
LPD	:	Liters per day
MFAL	:	Marginal Farmers and Agricultural Labourers.
MPCS	:	Milk Producers Cooperative Society
NCA	:	National Commission on Agriculture
NDDB	:	National Dairy Development Board
OF	:	Operation Flood

OMFED	:	Orissa State Cooperative Milk Producer's and Marketing Federation.
OWDP	:	Orissa Women Dairy Project
SFDA	:	Small Farmers Development Agency
STEP	:	Support to Training and Employment Programme
WDCLP	:	Women's Dairy Cooperative Leadership Development Programme
WDCS	:	Women Dairy Cooperative Society

List of Tables

List of Figures & Plates

PLATES

Chapter - 1

Dairy Development Scenario World, India & Odisha

Livestock Population and it's Economic Contribution

According to 18th Livestock census of India (2007), India has about 199.1 million cattle, which constitute 44.29 per cent of Asian and 12.7 per cent of the world cattle population. The current buffalo population is 105.34 million which accounts 57 per cent of Asian and 55 per cent of the world buffalos. The major output of the entire livestock sector is from the milk, which is the second most agricultural commodities next to paddy. According to the annual report of the Ministry of Agriculture and Cooperation, during 2006-07, the contribution of milk alone (Rs.1,44,386 crore) was higher than paddy (Rs.85,032 crore), wheat (Rs.66,791 crore) and sugarcane (Rs.28,488 crore). The value of output from meat group animals as per estimates of Central Statistical Organization (CSO) at current prices was Rs. 34, 310 crore during 2006-07. Total export earnings from livestock, poultry and related products were Rs.5,213.80 crore during 2006-07. Eighty percent rural households keep livestock for their subsidiary income, important for employment, livelihood, nutrition, constant flow of quick and fast income, reduces the vulnerability of agricultural production. Trends indicate that animal protein consumption would rise proportionately with increase in income. The 18th Livestock Census (2007) has placed the total livestock population at 529.698 million and total of poultry birds at 648.830 million.

Table 1.1 : Livestok population in India (2007)

Cattle	199.1
Adult Female Cattle	73.0
Buffalo	105.3
Adult Female Buffalo	54.5
Total Bovines	304.4
Sheep	71.6
Goat	140.5
Horses and Ponies	0.6
Cammels	0.5
Pigs	11.1
Mules	0.1
Donkeys	0.4
Yak	0.1
Mithun	0.3
Total live stock	529.7
Poultry	648.8

Table 1.2 : Milk production by species between India & world

Species	India	World
Buffalo Milk	56%	11.6%
Cow Milk	40.26%	85%
Others	4%	3.4%

Global Milk Production

The latest FAO (Food Agricultural Organisation of United Nations) outlook suggests that world milk production will increase by 14 million tonnes (2%) to 723.8 million tonnes in 2011. Most of this increase is expected to come from developing countries such as Brazil, China and India. The same report also indicates that cheese is growing faster than other dairy products, with world exports reaching 2.3 million tones in 2010, 6% higher than 2009. US milk production increased by 2.01% between January and April compared with the same period last year. New Zealand is up by 4% for the 2010/11 season and Austrialia finished their season slightly ahead of last year. Argentina's milk production is still increasing and is up by almost

17% for the January / April period from the same period in 2010. Argentina in 2010 produced over 11 million tonnes. Brazil, which produced 30.84 million tonnes in 2010, is up by 7% which amounts to 20.16 million tonnes. In EU there are 17 countries showing decreases. The five largest milk producing countries Germany, France, UK, Netherlands, and Italy produced 7,926 million tonnes (7.7 bn litres) in March 2011, which was 3.7% more than the previous year.

Table 1.3 : Milk Production by top countries in the World

Country	Total Milk Production (Tonnes)
India	112,114,290
United States	85,859,400
China	39,946,410
Pakistan	34,362,000
Russia	32,561,674
Brazil	29,255,768
France	24,217,730
Newzeland	15,400,000
United Kingdom	13,236,500
Italy	13,063,200
Turkey	12,542,172

World Milk Production by Species of Animals

Cow contributes maximum milk that is 553.0 million tones as against 82.00 million tones from buffaloes,12.50 million tones from goats, 8.6 million tones from sheep milk and 1.3 million tones from others to the total milk production of the world. Further, the cow milk production is maximum in the world, thus people of the western countries prefer to rear cows than buffalos for milk production.

Table 1.4 : World Milk Production by species of animals (million tones)

Million Tonnes	2003	2004	2005	2006	2007
Cow Milk	516.7	521.3	532.5	543.2	553.0
Buffalo Milk	75.5	76.5	78.6	80.5	82.0
Goat Milk	12.5	12.5	12.4	12.5	12.5
Sheep Milk	8.3	8.6	8.6	8.6	8.6
Others	1.3	1.3	1.3	1.3	1.3
All Milk	614.3	620.2	633.5	645.0	655.0

Source : IDF Bulletin No. 368/2007

Cow Milk Production by World Region

As per IDP Bulletin Report (2007), world milk production is 655.00 million tones to which maximum share comes from European Union followed by Asian Continent, North America. Therefore, EU and the Asian continent has great potential of milk production.

Table 1.5 : Cow milk production by world region (million tonnes)

Million Tonnes	2003	2004	2005	2006	2007
Africa	44.7	45.6	46.1	46.3	46.5
North America	85.3	85.7	88.3	90.3	90.51
Central America	15.7	15.8	15.8	5.9	16.2
South America	46.5	46.7	48.3	49.5	50.5
Selected Asian Countries*	71.8	78.0	84.5	91.5	97.0
Asian CIS	12.5	12.8	13.1	13.2	13.3
EU-15	121.9	120.6	121.1	119.8	120.2
El-10	21.8	21.6	21.8	21.9	22.0
EU-25	143.7	142.3	143.0	141.7	142.2
EU-27	150.9	149.9	150	149.2	150.0
Other CEEC	3.2	3.6	3.8	3.7	3.8
European CIS inclu. Siberia	52.7	52.5	52.0	52.0	52.5
Other W. Europe**	5.6	5.6	5.6	5.6	56
Oceania***	25.0	24.5	25.0	25.6	25.0
World	516.7	521.3	532.5	543.2	553.0

** Only South Africa, Kenya, Zimbabwe* *** USA,Canada, Mexico*

**** China, Japan & Asian CIS*

Source : IDF Bulletin, 2007

Buffalo Milk Production

India stands first in buffalo milk production and Pakistan, Egypt, Italy occupies 2nd, 3rd and 4th position in the world respectively. Such outstanding contribution of India to the world milk production is because of availability of some of the finest buffalo breeds like Murrha, Mehesana, Jaffarabadi, Nagapuri, Surti, Nili-Ravi, Bhadwari etc. in the country.

Table 1.6 : Buffalo Milk Production

Million Tonnes	2003	2004	2005	2006
India	48.0	49.4	50.9	52.5
Pakistan	18.6	19.2	19.7	20.0
Egypt	2.1	2.2	2.2	-
Italy	0.1	0.1	0.1	-
World	75.5	76.5	78.6	80.5

Source : IDF Bulletin No. 368/2007

Future Production & Consumption Trends

The United Nations (UN) has projected 3 scenarios for our future global population according to the fertility rates. With the ever increasing global population, there comes a greater need for increased livestock production to feed the world. With data from United States Department of Agriculture (USDA) and Food and Agricultural Policy Research Institute (FAPRI), livestock production is projected to increase from 220 million metric tons in 2008 to 250 million metric tons in 2018.

There are projections that demand for foodgrains would increase from 192 million tonnes in 2000 to 345 millon tonnes in 2030. Hence in the next 20 years, production of foodgrains needs to be increased at the rate of 5.5 million tonnes annually. The demand for high-value commodities (such as horticulture, dairy, livestock and fish) is increasing faster than foodgrains- for most of the high-value food commodities demand is expected to increase by more than 100% from 2000 to 2030.

Diary Development in India

Diary development has been given adequate place in the national agricultural development plan in India. Since then, both productivity and production of livestock including milk has been increased significantly over the plan periods of 50 years. The milk production in India has been somewhat slow from 1951 to 1970 depicting an increase of about 5 million tones only during a period of 20 years but has registered a quantum jump from 22.5 million tones in 1970-71 to 31.6 million tones in 1980-81, 53.9 million tones in 1990-91 and 76.87 million tones in 1998-99, thus making India the largest producer of milk in the world. The causes of such galloping rise in milk production during the past few decades had been mainly due to increase in the number of milch animals, particularly buffaloes, crossbred cows and improved veterinary services and animal husbandry practices (Diary Year Book,

2007). Dairying in India, today, has emerged as an important sub-sector accounting for nearly two-thirds of the total livestock contribution to GNP with an encouraging growth rate of 5 percent.

India ranks first in world milk production, its production having increased from 17 million tones in 1950-51 to 112.00 million tonnes by 2009-10. At present India stands first in the world in total fluid milk production followed by USA, China, Pakistan and Russia. The per capita availability of milk has increased from 112 grams per day in 1968-69 to 263 grams per day in 2009-10, but is still low compared to the recommendations of ICMR of 285 grams per day. About 80 per cent of milk produced in the country is handled in the unorganized sector and the remaining 20 per cent is equally shared by cooperatives and private dairies. Over 1.33 lakh village level dairy cooperative societies spread over 265 districts in the country, collect about 25.1 million litres of milk per day and market about 20 million litres. The efforts of the Government in the dairy sector are concentrated in promotion of dairy activities in non-Operation Flood areas with emphasis on building cooperative infrastructure, revitalization of sick dairy cooperatives and federations and creation of infrastructure in the states.

Table 1.7 : Production and per capita availability of milk

Year	Per capita availability (grams/day)	Milk production (million tonnes)
1990-91	176	53.90
2000-01	220	80.60
2005-06	241	97.10
2006-07	246	100.90
2007-08	252	104.80
2008-09	258	108.50
2009-10	263	112.00

Milk Production from 1950 to 2020

1950	-	17 million tonnes
1996	-	70.8 million tonnes
1997	-	74.3 million tonnes
(Projected) 2020	-	240 million tonnes

Table 1.8 : Production of major livestock products and fish

Year	Milk (Million tonnes)	Eggs (Million Nos.)	Fish (Thousand tonnes)
1950-51	17.0	1832	752
1960-61	20.0	2881	1160
1970-71	22.0	6172	1756
1980-81	31.6	10060	2442
1981-82	34.3	10876	2444
1982-83	35.8	11454	2367
1983-84	38.8	12792	2506
1984-85	41.5	14252	2801
1985-86	44.0	16128	2876
1986-87	46.1	17310	2942
1987-88	46.7	17795	2959
1988-89	48.4	18980	3152
1989-90	51.4	20204	3677
1990-91	53.9	21101	3836
1991-92	55.7	21983	4157
1992-93	58.0	22929	4365
1993-94	60.6	24167	4644
1994-95	63.8	25975	4789
1995-96	66.2	27198	4949
1996-97	69.1	27496	5348
1997-98	72.1	28689	5388
1998-99	75.4	29476	5298
1999-00	78.3	30447	5675
2000-01	80.6	36632	5656
2001-02	84.4	38729	5956
2002-03	86.2	39823	6200
2003-04	88.1	40403	6399
2004-05	90.7	45201	6304
2005-06	97.1	46231	6572
2006-07	100.9	50653	6869
2007-08	104.8	53532	7127
2008-09	108.5	55638	7608P.

It is observed that during 2009-10 the milk production in our country is112 million tones, considering a growth rate of 4 per cent in milk production (Indian Dairy Man, 2007). Assuming a compound growth rate of 1.86 per cent (1990-91 to 1998-99) of human population in India (CSO, 1999) the per capita availability of milk is projected to be 264 grams per day (96.3 kg/head/year at 4% annual growth rate in milk production). At this prevailing growth rate of human population and milk production,in the year 2007-08, India became self-sufficient to provide 250 gms of milk per day to it's population. However, to provide 280 gms milk per capita per day by 2011-12, a growth rate of 4.5 per cent per annum has to be achieved. By reducing population growth rate to 1.50 per cent per annum, the availability of milk would be 280 gms/ capita/day by the year 2011-2012.

Table 1.9 : Projected human population (crores), Milk production (million tones) and per capita availability (kg/year) during next decade in India

Year	Human popu -lation	Growth Rate in Milk production									
		4%		4.5%		5%		5.5%		6.0%	
		A	B	A	B	A	B	A	B	A	B
1999-00	99.31	77.69	78.23	78.06	78.60	78.43	78.97	78.81	79.36	79.18	79.73
2000-01	101.16	80.80	79.87	81.55	80.63	82.36	81.42	83.14	82.19	83.93	82.97
2001-02	103.04	84.03	81.55	85.25	82.73	86.47	83.92	87.72	85.13	88.97	86.35
2002-03	104.96	87.39	83.26	89.08	84.87	90.80	86.51	92.54	88.17	94.31	89.85
2003-04	106.91	90.88	85.01	93.09	87.07	95.34	89.18	97.63	91.32	99.97	93.51
2004-05	108.90	94.52	86.80	97.28	89.33	100.11	91.93	103.00	94.58	105.96	97.30
2005-06	110.92	98.30	88.62	101.66	91.65	105.11	94.76	108.66	97.96	112.32	101.26
2006-07	112.98	102.23	90.49	106.23	94.03	110.37	97.69	114.64	101.47	119.06	105.38
2007-08	115.09	106.32	92.38	11.01	96.45	115.88	100.69	120.95	105.09	126.20	109.65
2008-09	117.23	110.57	94.32	116.01	98.96	121.68	103.80	127.60	108.85	133.78	114.12
2009-10	119.41	115.00	96.31	121.23	101.52	127.52	106.99	134.62	112.74	141.80	118.75

(A-projected milk yield, B- Availability of milk per capita per year.)

India dairy industry has to compete globally due to WTO sanctions and price of raw milk paid to the producers. Although the productivity of our animals is very low as compared to the animals of developed countries, yet low input costs have resulted in low price of raw milk which resulted trade competitiveness. The average milk price paid to producers in India is Rs. 8.08 to that of Rs, 7.96 and Rs, 7.24 per kg in Australia and New Zealand respectively during 1999 (IDF -2000). However, the price paid per kg of milk to the producers was much higher in European countries (Rs.13.54), USA (Rs. 13.95) Japan (Rs.31.58) respectively.

Dairy Development in Odisha

Odisha is the 10th largest state in India, has an area of 15.57 million hectares and human population of 36.80 million (2001). Livestock production in Odisha is the endeavour of the small holders and takes place in millions of tiny holdings scattered across the length and breadth of the state. In Odisha, marginal and small farmers together account for 80 per cent and they own only 47% of the land, 80% of milch animals, 55% of drought animals, over 73% of poultry and almost the entire sheep, goat and pig population. They constitute poorer sections of the society in terms of landless, marginal and small holder categories.

Table 1.10 : Yearwise production and per capita availaibility of milk

Year	Production of milk (1000 mt.)	Per capita availability of milk (gm per day)
2003-04	995.08	74
2004-05	1282.76	94
2005-06	1342.30	98
2006-07	1424.64	103
2007-08	1620.44	115
2008-09	1598.05	113
2009-10	1651.33	109
2010-11 (P)	1670.00	109

Source : Directorate of Animal Husbandry & Vetrinary Services Odisha.

Odisha took the first step towards dairy development in the year 1963-64 i.e. during third five year plan (1961-66), when the first liquid milk pasteurization plant of 6000 LPD was established at Fulnakhara under "Cuttack Milk Union Scheme". Subsequently, this milk plant has strengthened by the addition of small chilling units of 2000 LPD capacity at Niali and Nimapara in the district of Cuttack and Puri respectively. In 1968 a Pilot milk scheme was established at Kaimati in Dhenkanal district having a 500 LPD capacity chilling plant. Another pilot scheme was started in October 1971 at Berhampur with 1800 LPD.

The Intensive Cattle Development Project (ICDP) was started in Cuttack district in September 1965, Berhampur in March 1972 and in Sambalpur district in February 1974. A Hill Cattle Development Project (HCDP) was started in Koraput under Indo-Danish collaboration in 1974.

It was during the middle of the fifth plan (1974-79) 12 District Cooperative Milk Unions were formed. Proposals for a Dairy Development Corporation came up in 1977, but it did not materialize. The cattle development activities continued by Animal Husbandry Department and other agencies like MFAL, SFDA, IRDP, DPAP, DRDA and BAIF.

Operation Flood Programmes

Odisha was not included in operation Flood-I, which was launched in July 1970 as because the NCA did not find it agro-climatically and demographically suitable for milk production. However, on the subsequent request of Govt. of Odisha to NDDB, the Milk sheds comprising four undivided districts like Cuttack, Puri, Dhenkanal and Keonjhar were included and an agreement was drawn up on 30.05.1979 between Indian Diary Corporation and Govt. of Odisha for implementation of Operation Flood –II Project. The Odisha State Cooperative Milk Producers' Federation Ltd. (OMFED) a three tire structure was registered on 28.01.1980 after which dairy development in Odisha took a momentum. With the launching of operation Flood in Odisha, the State's only milk plant at Fulnakhara, Cuttack was taken over by OMFED on 26.01.1981 and expended to 15,000 LPD capacities. Subsequently Bhubaneswar Dairy of 60,000 LPD was commissioned in December 1985, Dhenkanal and Keonjhar chilling plants in September 1984 and January 1985 respectively. As on Sept.'87, 441 Dairy Cooperatives were organized with a membership of 22,668 farmer members with a daily average milk procurement of 28,500 Kgs per day.

The operation Flood III programme was implemented in the said district milk unions from Oct. 87. Sambalpur district was included in August. 89. Expansion of Bhubaneswar Dairy and Chilling Plant (Tirtol) and new chilling plants (Nimapara) were made. Few other agencies have been involved for dairy development in the state. They are;

Indo-Swiss Project, Odisha (ISPO)

ISPO was established in 1991 under a bilateral agreement between the Govt. of India and the Govt. of Switzerland. The overall objective of the project is to contribute to increased productivity of the farm households and livestock. It is concentrating its activity in Ganjam and Gajapati District of the State.

Integrated Livestock Development Project (ILDP), Koraput

The ILDP is a bilateral collaboration project under an international agreement between the Govt. of Denmark and Odisha. The project partners are DANIDA & Odisha State Deptt. of Animal Husbandry. The Project started its operations in 1993 in Koraput district & primarily aims at tribal development with increased awareness and better skills in animal husbandry & fodder production.

State Technology Mission on Dairy Development(STMDD)

The Govt. of Odisha has set up a "State Technology Mission for Dairy Development". The aim is to bring all dairy development activities in the state under a unified control. The mission will cover the integrated dairy development programmes funded by Govt. of India, State Plan schemes for dairy development, operation flood infrastructure and all funds for dairy development available to the state from all sources. The Managing Director, OMFED is the member convener & Mission Director and function under the guidance & control of an Empowered Committee.

The Utkal Gomangal Samiti (UGS)

It's main objective is for upgrading of indigenous cattle and buffalo through natural service. It's activities supplement the efforts of the Deptt. where A. 1 facility have not reached so far.

Department of Veterinary Services and Animal Husbandry

It is the major provider of breeding services for cattle & buffalo in the state. It has 2939 nos. of A.I centre and 540 dispensaries & hospitals in 30 districts of the state. The frozen semen production stations were expanded and modernized with the assistance from Technology Mission on Dairy Development (TMDD). Dairy Development in non-operation flood districts through IDDP funds is also undertaken by Animal Husbandry Department.

OMFED and its Cooperative Network

As on date there are 319 A.I. centre functional in its operational area. Primarily as a production support services for their producer members.

Women Dairy Project (WDP), Odisha

Under Support to Training & Employment Programme for Women (STEP), the OMFED has got the sanctions of Rs. 20.58 crore for implementation of women dairy cooperatives in the undivided districts of Cuttack, Puri, Dhenkanal, Sambalpur and Keonjhar through its district Milk Unions. It has so far organized 918 all women dairy coop. & enrolled 68.851 women members since 1995 and contributes 1/3rd of total milk procurement.

Milk Production

Odisha occupies 17th position in milk production in the country. Milk production in Odisha involves three species of milch animals like cattle, buffaloes and goat. Buffaloes contribute 20%, cross-bred cows contribute 33%, local cows contribute 47% to the total milk production of the state respectively. However, contribution of goat to the total milk production is negligible.

Milk production in Odisha grew from 3.16 lakh mt. and 13.42lmt in the year 2005-06 (lmt) in 1970-81 to 6.58 lmt. in 96-97 with 5.08 % growth per annum. 20 % of the same (1.32 lmt.) is consumed in producer households and rest is traded in the State, bulk of it is for domestic consumption and some 1200mt goes to Calcutta in the form of Chhena. OMFED's entire milk procurement is traded entirely in domestic market except in flush when some quantity is sent to Mother Dairy, Calcutta.

In 2003, Odisha had a total livestock population of 24.02 million of which the share of cattle and buffalo population is 15.72 millions. However, 2007 census shows there is decrease of cattle & buffalow population as shown in table below:

Table 1.11 : Cattle and Buffalo population 2007 Census. (in lakh)

Category	All Total		Female		Milch		Breedable	
	Total	CB	Total	CB	Total	CB	Total	CB
1	2	3	4	5	6	7	8	9
Cattle	123.06	17.03	58.53	8.25	27.1	3.32	31.72	4.47
Buffalo	11.9	NA	5.88	NA	2.81	NA	3.35	NA
Total	134.99	NA	64.41	NA	29.91	NA	35.07	NA

Source : Directorate of Animal Husbandry & Veterinary Services, Odisha

In Odisha as such there are no recognized breeds of cattle native to it, but to it's credit this state does have some local strains who have some distinct name in the concerned locality. Among these local strains of cattle Binjharpuri, Ghumusari, Khariar, Mottu are famous in the state. The cross-bred cattle account for only about 5.30 per cent (0.74 million) of the total cattle population. The cross–bred cows being small in number contribute around 33% of milk to the total milk production of the state. However, average milk yield of the indigenous nondescript cattle is only 166 kg per lactation of 100 days and that of cross-bred cattle is 1167 kg per lactation of 300 days. Buffalo population in the state is 1.65 million sharing 6.81 per cent of the total livestock population. There are good number of local strains of buffaloes but none of them are recognized. These are Paralakhemundi, Manda, Jirangi, Sambalpuri, Dhenkanal, Paradeep, Kujanga, Beda, Chilika and Kalahandi. They produce on an average 4-5 kg milk per day. Being such small numbers they still contribute about 20% of the total milk production of the state.

Among all the districts, Cuttack district heads in milk production producing around 68.8 thousand metric tones of milk followed by Puri, Ganjam, Jagatsinghpur, Khurda. However, in the per capita milk consumption scenario Puri district is well ahead of other districts consuming about 130 gm milk per day followed by Jagatsinghpur 113 gm, Cuttack 88gm, Baragarh 66gm, and Deogarha 61 gm of milk/head/day.

However, as per ICMR recommendation, the per capita requirement of milk per day should be 285 gm. But the present position of availability of milk in Odisha is far below i.e. 109 gm/day/head, while the situation of milk availability in other states in India are more comfortable and at an appreciable level i.e. in Punjab it is 794 gm. Haryana 592 gm. Himanchal Pradesh 339 gm, Rajasthan 319 gm and Gujarat 259 gm/day per individual.

There has been a quantum progress in the dairy sector in Odisha and to give further momentum to this venture in future, an objective has been fixed to increase per capita consumption of milk at per with national per capital availability of 285 gms/day which is at present 109 gms/day in our state.

Dairy development in the state is presently taken up by OMFED through Operation Flood programmes in 14 districts. These districts are Cuttack, Jajpur, Kendrapara, Jagatsinghpur, Puri, Khurda, Nayagarh, Dhenkanal, Angul, Keonjhar, Sambalpur, Baragarh, Jharsuguda, Deogarh. Similarly non-Operation Flood programmes are implemented in other 14 districts by A.H. Department. These districts are; Koraput, Rayagada, Malkanagiri, Nuapada, Nawarangpour, Phulbani, Sundargarh, Balangiri, Kahahandi, Boudh, Sonepur,

Balasore, Bhadrak and Mayurbhanj. In rest two districts dairy development programmes is implemented by ISPO.

Table 1.12 : Achievements of OMFED

Milk procurement	-	3.7 lakhs LPD
Processing plant	-	6
Bhubaneswar	-	1,25,000LPD
Rourkela	-	30,000 LPD
Sambalpur	-	30,000 LPD
Dhenkanal	-	10,000 LPD
Keonjhar	-	10,000 LPD
Berhampur	-	8,000 LPD
Chilling Plant	-	17
Milk Societies	-	5,022
Enrolled members	-	1,12,000
Women Dairy Project	-	15 districts
(Cuttack, Puri, Dhenkanal, Sambalpur, Keonjhar, Ganjam, Bargada, Jharsuguda, Deogarh, Jagatsinghpur, Kendrapara, Nayagarh, Khurda, Anugul and Gajapati)		
Support by Ministry of Women and Child		
Welfare, Govt. of India	-	All complete
Project period (Keonjhar & Ganjam)	-	3 years
Total Women Dairy Societies		
Projected	-	950
Achievements	-	918
Enrolled members target	-	71,000
Achievements	-	68,851
Milk production target	-	4,18,955 LPD
Achievements	-	3,70,179 LPD
Women A.I. Centre		
Target	-	165
Achievements	-	161

Strategies for Future

- Setup of additional dairy plant with capacity of 1 lakh LPD near Bhubaneswar with an estimated cost of Rs.6 crores. Sources of funds DRDA/NDDB/Bank.

- Marketing infrastructure near Kolkatta.
- Setting up an automatic bottling plant at Sambalpur in the campus of OMFED food processing centre. In future mineral water is likely to be manufactured in this plant.
- Setting up a milk powder plant for OMFED at Bhubaneswar Dairy.
- By 2005 there would be a surplus of milk by 1.0 lakhs lts per day. Hence there is a proposal for a powder plant with an estimated cost Rs,.7.5 crores of on 75:25 basis by GOI and OMFED.

A.H. Department (IDDP) (Non-Operation Flood Districts)

MPC's	-	544
Chilling Plants (Capacity - 40,000 LPD)	-	18
Dairy Processing Plants (Capacity 20,000 LPD)	-	5
(Bolangir, Bhawanipatna, Boudh, Koraput,		
Nawarangpur) Membcrs	-	32,166
Milk Procurement (Target-40,000 LPD)	-	17,000 LPD
ISPO Now taken over by OMFED		
MPC's	-	128
Chilling Plants (Capacity 2,000 LPD)	-	2
(Aska, Parlakhemundi)		
Dairy Processing Plant(Capacity - 8,000 LPD)	-	1

Ninth Plan Budget and Constraints

- We could not achieve the target of 40,000 ltrs per day due to low funding by GOI. Against the 9th plan outlay of Rs.6.53 crores only Rs. 2.93 crores has so far been received. 10th Plan projections
- Improvement of indigenous cattle and Buffalo to boost milk production in thrust areas.
- Enrollment of more members.
- Formation of primary milk producers' societies in areas where there is no MPCS.
- Increasing the strength of chilling plants and dairy processing plants.

- The total milk procurement to be increased from 1.37 lakhs LPD to 3.0 lakhs LPD.

Status and Trend of Milk Availability in Odisha

Percapita availability of milk in Odisha is very low when compared to national percapita availability. Starting from a meagre availability of milk that too 33 g during the year 1982-83 has increased at a slow pace to the level of 109 g in the year 2009-10 which is far short of the level of requirement i.e., 285 g as recommended by ICMR. There has been a deficit of 176 g. If the past trend of growth continues, by 2010 the state would be able to make availability of milk 220 g percapita per day.

Table 1.13 : Per capita availability of milk in different years

Years	Per - Capita availibility
1	2
1982-83	33gm
1983-84	34 gm
1984-85	37 gm
1985-86	38 gm
1986-87	38 gm
1987-88	39 gm
1988-89	40 gm
1989-90	40 gm
1990-91	41 gm
1991-92	43 gm
1992-93	45 gm
1993-94	46 gm
1994-95	47 gm
1995-96	49 gm
1996-97	50 gm
1997-98	52 gm
1998-99	53 gm
1999-00	55 gm
2000-01	67 gm
2001-02	70 gm
2002-03	71 gm
2003-04	74 gm

Contd.

2004-05	94 gm
2005-06	98 gm
2006-07	103 gm
2007-08	105 gm
2008-09	113 gm
2009-10	109 gm
2010-11 (P)	109 gm

❑❑❑

Chapter - 2

Women in Dairy Development

Status of Women

As per 2011 Census, women account for 586.46 million, and represent 48.5% of India's total population. However, from global perspectives, they constitute half of global population, and 1/3rd labour force, but receive only 1/10th of world income, and own less than 1% of world's property. Further from national point of view as Reddy *et al.* (1994) in his study opined that "although women constitute 50 percent of India's population, perform two-thirds of the work and produce 50 percent of food commodities consumed by the country and they earn only one-third of the remuneration and 10 percent of the property or wealth of the country.

The vast majority of working women are uneducated cultivators and agricultural labourers residing in villages. Most of the women, working on the family farm are unpaid labourers. 63 percent of all educated women with at least a high school certificate are unemployed, and of the 37 percent who are employed 15 percent are clerks, 5 percent teachers, 7 percent nurses and 10 percent are other professionals. Only 6 percent of Government employees are women (Jain, A.K. 2004). Rural Women form the most important productive work force in the economy of majority of the developing nations including India. Agriculture, the single largest production endeavour in India, contributing about 18% of GDP, is increasingly becoming a female activity. Agriculture sector employs 4/5th of all economically active women in the country. 48% of India's self-employed farmers are women. There are 75 million women engaged in dairying as against 15 million men and 20 million in animal husbandry as compared to 1.5 million men.

Despite progress in several key indicators, a gender analysis demonstrates that women continue to be relatively disadvantaged in matters of survival, health, nutrition, literacy and productivity. Further, more than 90 percent of rural women are unskilled, restricting them to low paid occupations. Women generally have no control over land and other productive assets which largely exclude them from access to institutional credit. Further, agricultural wages for women, on an average 30-50 percent less than those for men. Though they work for longer hours under inhospitable conditions, their participation in decision making process is less than desirable.

Govt. of India had taken some of the welfare measures to provide security of participation in different Govt. programmes and generate their own income. In that context the programmes like CSWB (1953), declaring 1975-85 as women's decade, ICDS (1975-76, DWCRA (1982), Department of Women and Child development (1985), National Commission for Women (1992), declaring 2001 as women's empowerment year and Rashtriya Mahila Samrudhi Yojana (2003).
A shift in the approach from welfare to development of women took place in Sixth Five Year Plan. The Eighth plan (1992-97), with human development as its major focus, played a very important role in the development of women.

In addition to the above, there are number of programmes, for improving the status of women. The Indian Mahila Jojana (IMY, launched in 1995-96 has been retitled as Swayamsiddha in 2001 to empower women through awareness generation, achievement of economic strength through micro-level income generating activities and establish conveyance of various services (literacy, health, rural development etc.). For women the literacy rate stands at 54.16 per cent. Still, 245 million Indian women cannot read or write, comprising the world's largest number of unlettered women. National averages in literacy conceal wide disparities. For instance, while 95 per cent of women in Mizoram are literate, only 34 per cent of women in Bihar can read and write. The average Indian female has only 1.2 years of schooling, while the Indian male spends 3.5 years in school. More than 50 per cent girls drop out by the time they are in middle school. Similarly, life expectancy has increased for both the sexes; it has increased to 64.9 years for women and 63 years for men according to UN Statistic Division (2000). The Working women population has risen from 13% in 1987 to 25% in 2001. However, the UNFPA State of World Population 2005 states that about70% of graduate Indian women are unemployed. Women constitute 90 per cent of the total marginal workers of the country. Rural women engaged in agriculture form 78 per cent of all women in regular work. They are a third of all workers on the land. The traditional gender division of labour ensures that these women get on average 30 per cent lower wages than men. The total employment of women in organized sector is only 4 per cent. A study by SEWA of 14 trades found that 85 per cent of women earned only 50 per cent of the official poverty level income.

Genesis of Cooperatives in India

Cooperation dates back as far as human beings have been organising for mutual benefit. In 1761, the Fenwick Weavers' Society was formed in Fenwick, East Ayrshire, Scotland to sell discounted oatmeal to local workers. Its services expanded to include assistance with savings and loans, emigration and education. In 1810, Welsh social reformer Robert Owen, from Newtown in mid-Wales, and his partners purchased New Lanark mill from Owen's father-in-law David Dale and proceeded to introduce better labor standards including discounted retail shops where profits were passed on to his employees. Owen left New Lanark to pursue other forms of co-operative organization and develop co-op ideas through writing and lecture. Co-operative communities were set up in Glasgow, Indiana and Hampshire, although ultimately unsuccessful. In 1828, William King set up a newspaper, The Cooperator, to promote Owen's thinking, having already set up a co-operative store in Brighton. The Rochdale Society of Equitable Pioneers, founded in 1844, is usually considered the first successful co-operative enterprise, used as a model for modern co-ops, following the 'Rochdale Principles'. A group of 28 weavers and other artisans in Rochdale, England set up the society to open their own store selling food items they could not otherwise afford. Within ten years there were over 1,000 co-operative societies in the United Kingdom.

Concept of Cooperative

Cooperatives as Legal Entities

A cooperative is a legal entity owned and democratically controlled by its members. Members often have a close association with the enterprise as producers or consumers of its products or services, or as its employees. Cooperatives may take the form of companies limited by shares or by guarantee, partnerships or unincorporated associations.

Aspects of Economic Rationale

Cooperatives often share their earnings with the membership as dividends, which are divided among the members according to their participation in the enterprise, such as patronage, instead of according to the value of their capital shareholdings.

Identity

Cooperatives are based on the cooperative values of "self-help, self-responsibility, democracy and equality, equity and solidarity".

Cooperative Principles

Voluntary and Open Membership

Democratic Member Control

Member Economic Participation

Autonomy and Independence

Education, Training and Information

Cooperation among Cooperatives

Concern for Community

In the tradition of their founders, cooperative members believe in the ethical values of honesty, openness, social responsibility and caring for others. Such legal entities have a range of unique social characteristics. Membership is open, meaning that anyone who satisfies certain non-discriminatory conditions may join. Economic benefits are distributed proportionally according to each member's level of participation in the cooperative, for instance by a dividend on sales or purchases, rather than divided according to capital invested. Cooperatives may be generally classified as either consumer cooperatives or producer cooperatives. Cooperatives are closely related to collectives, which differ only in that profit-making or economic stability is placed secondary to adherence to social-justice principles. Gurumoorty (1995) opined that cooperation is the movement of people, by the people and for the people. Democracy is the principle of the cooperative movement. Cooperative institutions protect individual interests and pave the way for common interests. The services of the cooperative institutions improve the socio-economic conditions of the members and strengthen the economy of the country.

According to Oxford Advanced Learners' Dictionary, the word "Cooperative" means working together with another or others for a common purpose. The cooperatives involved in commerce or business are defined as those societies usually owned and run by the people involved with profits shared by them. Nainta (1995) defined Cooperative societies as autonomous democratic institutions for the welfare of the public. They are governed by the principles of cooperation, justice and quality. Whether, a primary, secondary or apex society, they have to work under specified framework i.e. Act, Rules and By-Laws framed by respective society for the management.

History of Dairy Development in India

The history of the dairy development in India is broadly classified into two distinct phases: pre- and post-Operation Flood. The earliest attempts of dairy development is traced back to British rule, when the Defence Department established military dairy farms to ensure the supply of milk and butter to the colonial army. The first dairy farm was set up in Allahabad in 1913 and subsequent facilities were established at Bangalore, Ootacamund and Karnal. These farms were well maintained and, even in the early stages, improved milch animals were raised. As animals were reared under farm conditions, some herd improvement was made using artificial insemination. This approach did not have any impact on the supply of milk to urban consumers, which was of major concern to civilian authorities but less important to the military.

With the growth of the population in urban areas, consumers had to depend on milk vendors who kept cattle in these areas and sold their milk, often door-to-door. As a result, several cattle sheds came into existence in different cities. This was not an environmentally sound approach. As the main objective of the milk vendors was to maximize profit, they started increasing the lactation period by using the focus system. In the process, these high-yielding cattle developed sterility problems, which considerably reduced the number of cavings. Once the cattle became unproductive, they were sold to slaughter houses. This practice systematically drained the country of its genetically superior breeds.

To some extent, the Second World War gave impetus to private dairies with modestly modernized processing facilities. In the cities of Bombay, Calcutta, Madras and Delhi, and even in some large townships, processed milk, table butter and ice-cream were available, though not on a large scale. Polsons, Keventers and the Express Dairy were some of the pioneer urban processing dairies. These dairies were not concerned with improving the breed of milch animals reared in rural pockets but instead were content with contracting milk supplies through middlemen or their own staff. Milk producers as well as consumers were exploited. These early modern systems did not bring about significant shifts in milk production, nor did they develop quality milch animals. To a large extent, despite modernized processing facilities, dairying remained unorganized.

Milk procurement from the rural areas and its marketing in the urban areas was the major problem in Indian dairying at the time India gained independence. In one of the earliest urban milk supply schemes, Polsons - a private dairy at Anand - procured milk from milk producers through middlemen, processed it and then sent the milk to Bombay, some 425 km away. Bombay was a good market for milk and Polsons profited immensely. In the mid-1940s, when the milk producers in Kaira asked for a proportionate share of the trade margins, they were denied even a modest increase. The milk producers went on strike, refusing to supply milk to Polsons. On the advice of Sardar Vallabhbhai Patel, a leader in India's independence movement, the milk producers registered the Kaira District Cooperative Milk Producers' Union, now popularly known as AMUL, in 1946. The Kaira union procured milk from affiliated village-level milk societies. This was the genesis of organized milk marketing in India, a pioneering effort that opened a new vista for dairy development in the country.

Between 1946 and 1952, AMUL's policy was directed towards obtaining monopoly rights for the sale of milk to the Bombay milk scheme. In 1952, it succeeded in achieving its purpose after the Government of Bombay cancelled the contract with Polsons and handed over the entire business of supplying milk from the Kaira district to AMUL. However, as the Bombay milk scheme was committed to purchasing all the milk produced by the Aarey Milk Colony in Bombay, it would not take AMUL's milk during the peak winter months. The disposal of this surplus milk posed difficulties for AMUL, forcing it to cut down on purchases from its member societies, which affected members' confidence. The answer was the production of milk products. In 1955, a new dairy plant was set up at Anand to produce butter, ghee and milk powder.

A second dairy was built in 1965, and a product manufacturing unit was established in 1971 to cope with increasing milk procurement. In 1993, a fully automatic modern dairy was constructed adjacent to the original AMUL dairy plant at Anand.

With the initiation of India's first Five-Year Plan in 1951, modernization of the dairy industry became a priority for the government. The goal was to provide hygienic milk to the country's growing urban population. Initial government action in this regard consisted of organizing "milk schemes" in large cities. To stimulate milk production, the government implemented the Integrated Cattle Development Project (ICDP) and the Key Village Scheme (KVS), among other similar programmes. In the absence of a stable and remunerative market for milk producers, however, milk production remained more or less stagnant. During the two decades between 1951 and 1970, the growth rate in milk production was barely 1 percent per annum, while per caput milk consumption declined by an equivalent amount.

During the 1960s, various state governments tried out different strategies to develop dairying, including establishing dairies run by their own departments, setting up cattle colonies in urban areas and organizing milk schemes. Almost invariably, dairy processing plants were built in cities rather than in the milksheds where milk was produced. This urban orientation to milk production led to the establishment of cattle colonies in Bombay, Calcutta and Madras. These government projects had extreme difficulties in organizing rural milk procurement and running milk schemes economically, yet none concentrated on creating an organized system for procurement of milk, which was left to contractors and middlemen. Milk is perishable nature and relative scarcity gave the milk vendors leverage, which they used to get considerable advantage. This left government-run dairy plants to use large quantities of relatively cheap, commercially imported milk powder. The daily per caput availability of milk dropped to a mere 107 g during this time. High-fat buffalo milk was extended with imported milk powder to bring down the milk price, which resulted in a decline in domestic milk production. As the government dairies were meeting barely one-third of the urban demand, the queues of consumers became longer while the rural milk producer was left in the clutches of the trader and the moneylender.

All these factors combined left Indian dairying in a most unsatisfactory low-level equilibrium. The establishment and prevalence of cattle colonies emerged as a curse for dairying in the rural hinterland as it resulted in a major genetic drain on the rural milch animal population, which would never be replaced. City dairy colonies also contributed to environmental degradation, while the rural producer saw little reason to go for dairying.

AMUL formed the basis for the Anand Model of dairying. The basic unit in this model is the milk producers' cooperative society at the village level. These cooperatives are organizations of milk producers who wish to market their milk collectively. Membership is open to all who need the cooperative's services and who are willing to accept the responsibilities of being a member. Decisions are taken on the basis of one member exercising one vote. No privilege accrues to capital, and the economic returns, whether profit or loss, are divided among the members in proportion to patronage. Each cooperative is expected to carry out the continuing education of its members, elected leaders and employees. All the milk cooperatives in a district form a union that, ideally, has its own processing facilities.

All the unions in a state are normally members of a federation whose prime responsibility is the marketing of milk and milk products outside the state. There is also a fourth tier, the National Cooperative Dairy Federation of India (NCDFI), which is a national-level body that formulates policies and programmes designed to safeguard the interests of all milk producers. Each tier of the Anand organizational structure performs a unique function: procurement and services by the cooperative; processing by the union; marketing by the state federation; and advancing the interests of the cooperative dairy industry by the national federation. Thus, the Anand Model has evolved into an integrated approach to systematic dairy development.

National Dairy Development Board (NDDB)

In October 1964, on the occasion of the inauguration of AMUL's cattle feed plant, the then Prime Minister of India, Lal Bahadur Shastri, was impressed by the socio-economic changes brought about by the milk cooperatives, he expressed the desire for a national-level organization to replicate Anand Model dairy cooperatives throughout the country and to make available multidisciplinary, professional dairy expertise to dairies in the public and cooperative sectors. Thus, in 1965, NDDB was registered under the Societies Registration Act, the Charitable Trust Act and the Public Trust Act. Consistent with its model and mandate, NDDB headquarters were established at Anand.

Operation Flood (OF)

Operation Flood was a rural development programme started by India's National Dairy Development Board (NDDB) in 1970. One of the largest of its kind, the programme objective was to create a nationwide milk grid.It resulted in making India one of the largest producers of milk and milk products, and hence is also called the White Revolution of India. It also helped reduce malpractices by milk traders. Dr. Varghese

Kurien ,the architect of the White Revolution in India ,gave the professional management skills and necessary thrust to the cooperative.

The strategy for organized dairy development in India was actually conceived in the late 1960s, within a few years after the National Dairy Development Board (NDDB) was founded in 1965. It rested on the Operation Flood programme, which was conceived by the NDDB and endorsed by the government.

Operation Flood is a programme designed to develop dairying by replicating the Anand Model for dairy development, which has stood the test of time for almost half a century. The first phase of Operation Flood was launched in 1970 following an agreement with the World Food Programme, which undertook to provide as aid 1,26,000 tonnes of skim milk powder and 42,000 tonnes of butter oil to finance the programme.

The programme involved organizing dairy cooperatives at the village level; creating the physical and institutional infrastructure for milk procurement, processing, marketing and production enhancement services at the union level; and establishing dairies in India's major metropolitan centres. The main thrust was to set up dairy cooperatives in India's best milksheds, linking them with the four main cities of Bombay, Calcutta, Delhi and Madras, in which a commanding share of the milk market was to be captured. Operation Flood was implemented in three phases.

OF - Phase I

Traditionally, India has been an importer of dairy products. Phase I (1970–1980) was financed by the sale of skimmed milk powder and butter oil gifted by the European Union (then the European Economic Community) through the World Food Programme. NDDB planned the programme and negotiated the details of EEC assistance. In the begining of operation Flood-1 in 1970 certain set of aims were kept in view for the implementation of the programmers. Improvement by milk marketing was made by organizing dairy sector in the metropolitan cities including Bombay, Calcutta, Madras, and Delhi. The objectives of commanding share of milk market and speed up development of dairy animals respectively hinter lands of rural areas with a view to increase both production and procurement

During its first phase, Operation Flood linked 18 of India's premier milksheds with consumers of major metropolitan cities: Delhi, Mumbai, Kolkata and Chennai, thus establishing mother dairies in four metros.

OF - Phase II

Operation Flood Phase II (1981–1985) increased the milksheds from 18 to 136 and 290 urban markets expanded the outlets for milk distribution. By the end of 1985, a self-sustaining system of 43,000 village cooperatives with 4.25 million milk producers were covered. Domestic milk powder production increased from 22,000 tons in the pre-project year to 1,40,000 tons by 1989, all of the increase coming from dairies set up under Operation Flood. In this way EEC gifts and World Bank loan helped promote self-reliance. Direct marketing of milk by producers' cooperatives increased by several million litres a day.

Operation Flood was jointly sponsored by the European Economic Community, the World Bank, and India's National Dairy Development Board. The UNDP provided technical assistance by sending foreign experts, consultants, and equipment to India. The World Bank and its affiliates supported agricultural extension, social (community-based) forestry, agricultural credit, dairy development, horticulture, seed development, rain-fed fish farms, storage, marketing, and irrigation.

OF - Phase III

OF-Phase III (1985–1996) enabled dairy cooperatives to expand and strengthen the infrastructure required to procure and market increasing volumes of milk. Veterinary first-aid health care services, feed and artificial insemination services for cooperative members were extended, along with intensified member education.

Operation Flood's Phase III consolidated India's dairy cooperative movement, adding 30,000 new dairy cooperatives to the 42,000 existing societies organized during Phase II. Milksheds peaked to 173 in 1988-89 with the numbers of women members and Women's Dairy Cooperative Societies increasing significantly.

OF-Phase III gave increased emphasis to research and development in animal health and animal nutrition. Innovations like vaccine for Theileriosis, bypass protein feed and urea-molasses mineral blocks, all contributed to the enhanced productivity of milk animals.

Effects of the White Revolution

The white revolution gave a major boost to the dairy sector in India in the late 1960s by producing milk in rural areas through smallholder producer cooperatives and moving industrially- processed milk from these small holder sources to the urban demand centers, thus, establishing a much needed linkage between the rural producers and urban consumers. The United Nations has commended India's "White Revolution," saying a sharp increase in the production of milk has achieved twin goals of raising incomes of rural poor families and nutrition status of the people.

The White Revolution started in Gujarat and with its success over the years, other states have started implementing the concept of the White Revolution with a view to simultaneously improve and develop the economic and social parts of their rural sectors and linking them with their urban sectors.

After starting a White Revolution in the country and changing the way India consumes milk and its products, Amul is ready to start a second revolution. The cooperative milk giant is looking to double its milk production capacity by 2020. Years ago, the unique cooperative movement transformed India from an imports-dependent country to a self-sufficient one. Now, the demand for milk has increased with people's purchasing power going up and Amul is hoping to meet this demand with its capacity expansion.

GCMMF, which markets all Amul products, expects its sales to shoot up to Rs 27,000 crore in 2020 from an estimated Rs 6,700 crore in 2008-09. To achieve the 2020 target, Amul will increase the capacities of the existing 30 dairy plants and add some new ones

India's milk production rose from around 30 million tonnes in 1980 to 112 million tonnes by 2009-10 and despite increasing population, availability per person rose from less than 50 kilo calories per day in 1980 to 210 kilo calories per day in 2010, a report on hunger by the Food and Agriculture Organization (FAO) said. The report forecasts that India's dairy production will triple by 2020.

With government policies that facilitate rural credit and provide essential support services to promote milk production, the White Revolution will continue to play a significant role in reducing poverty and hunger.

FAO estimates that increasing milk production has boosted the incomes of 80 to 100 million families, the vast majority of whom are marginal or small farmers whose plots are often too small to support their families and landless laborers who depend on common grazing lands and forests for fodder.

Matching Demand and Supply

The white revolution carried on the extensive dairy development programs on the supply side, and this coupled with the increasing demand for value added milk products on the consumer side along with the country's population growth, increased urbanization and higher income, led to increased demand for milk. The various other factors contributing to the growth in milk production are given below:

- In the Indian diet, milk and milk products play a significant role.
- With a large lacto- vegetarian population, milk and dairy products are an important source of protein in diet.
- There is a perceptible shift towards the value- added food habits in which milk products form a large part.
- On the supply side, technological progress in the production and processing sectors, institutional factors and infrastructure played an important role.

Protection to Milk Industry

After the White revolution, smaller enterprises and cooperatives, with whom, milk processing and product manufacturing were mainly restricted, were given protection by means of high import duties, non-tariff barriers, restrictions on import and export and stringent licensing provisions. This also provided supplementary employment and income among the small and marginal farming households and landless wage earners.

Before the White revolution started, the farmer's income was derived almost entirely from seasonal crops. The income from milch buffaloes was undependable. Only private traders and middlemen gained as the marketing and distribution system was controlled by them. But currently, the scheme has ensured average returns of Rs. 1600 to Rs. 1700 per month, with the beneficiaries being mostly women. The Amul project seeks to bring poor households into the mainstream of economy by transforming dairying into active income generating enterprises. As many as 7500 below poverty line beneficiaries were targeted. The number has already been surpassed, and the project has already covered over 8000 beneficiaries.

All aspects of dairying have been exhaustively covered under the programme, in which innovations and modern technologies have been put to the best use for the rural population.

Women in White Revolution

Traditionally, India's dairy cooperative societies were run by men, but with time, this has gradually changed. Today, about 18% of cooperative members are women and many women cooperatives are functioning in the country. Over 1000 women in Gujarat's Sundarnagar district have proved that modern technologies can ameliorate the worries of poor rural women. Gujarat in fact, spearheaded the 'Operation Flood' movement in the 1970s, which sought to maximize milk production and profits through scientific techniques and attempted to free milk producers from the tyranny of the middlemen. Today, several NGOs are working to change this scenario to be more inclusive of women. The Self Employed Women's Association (SEWA), Ahmadabad, Gujarat was the first one to start the concept of all- women dairy cooperatives in Banaskantha district.

Initially, women were hesitant about depositing milk with the dairy. But through intensive training sessions, they began to see the benefits of associating with the dairy. On an average, a woman is able to deposit one to two litres of milk per day. The women associated with the dairy say that it is not the higher price they get for their milk that makes them keep coming back. There are other benefits, like bonus, payment every 10 days, information about fodder, mineral mixture and cattle health training. Bank accounts and transactions, as well as the day- to- day functioning of the cooperatives are operated entirely by women.

The women of these villages have developed an unshakeable faith in the machines. They are especially pleased with the measurement of fat content in milk. This helps them fix the price of the milk depending on its fat content, instead of the flat rate that the middlepersons gave them. They have also learnt techniques to ensure better health for their cattle, through de- worming and mineral mixture. This, in turn, leads to increased fat content in milk. Further, artificial insemination - which the women can afford because of the profits that the cooperatives earn - results in a better breed of cattle.

Thus, White revolution had a tremendous impact on the economic as well as the social life of the rural sector of India and had helped in the development of rural sector an established a linkage between the rural and the urban sector of India.

Women Dairy Cooperative Societies

Since the last decade or so the contribution of women in dairy production is gaining due reorganization. It is realized that women handle most of the critical jobs like bringing fodder from field, chaffing fodder preparing feed for animals, offering water to animals, protection of animals from ticks and lice, cleaning of animals and sheds, preparation of dung cakes, milking, ghee-making and marketing of produce. However, in many cases the income from dairy animals does not remain in the hands of the women and neither does the decision regarding sale and purchase. Similarly, despite plethora of programmes aimed at socio-economic emancipation in decision making process is not satisfactory at all levels. Realizing the potentiality of women in dairy development, some of the programmes like cooperatives development programme, women's dairy cooperative leadership development programme (WDCLP), and support to training an employment programme (STEP) for women with special reference to women dairy projects were launched to support women. CD programme in particular focused attention on women's leaders. Selected milk unions under CD programme had teams of five educator-trainers. Each team normally included a coordinator and four instructor. These teams conducted a variety of educational and training activities including mass educational programmes for women and general membership; training for management committee members, secretaries, and other dairy cooperative society's staff, and leadership development programme for women and youth (NDDB, 1994). Subsequently, NDDB launched women dairy cooperative leadership development programme (WDCLP) with an objective to increase women membership in WDCs. The aim of the WDCLP was to (a) take women beyond of their homes, (b) organize women into self-help groups (c) provide a platform to learn (d) develop their freedom by gently easing rigid social norms (NDDB, 1994). A number of dairy cooperative unions too started organizing all women dairy cooperatives in their milk sheds and such society was first organized in Kaira District, Cooperative Milk Producers Union, Gujarat in 1968 (Rao, 2000).

Dairy cooperative societies have been playing a pivotal role in bringing about white revolution in India. The first dairy cooperative society registered in Allahabad in UP in 1993 and was called "Katra Cooperative Dairy Society." The dairy cooperative movement in India has achieved a remarkable success in the past 50 years, more so in the last 20 years. As on March 2006 the dairy cooperative network includes

170 milk unions, which operates in over 346 districts, covers around 1,17,575 village level societies, is woned by around 12.4 milion farmer's members of which 3.2 millions were women. India's 3.9percent annual growth of milk production between 1995 -96 and 2005-06 surpassed the 2 percent growth in population; the net increase in availability was rount 2 percent per year. When comes to the marketing, in 2005-06, average daily cooperative milk marketing stood at 16806 lakh its annual growth has averaged about 5.8 percent compunded over the last five years. Dairy cooperatives now market milk in all metros, major cities and more than 800 towns/cities. During the last decades, the daily milk supply to each 1000 urban consumer has increased from 17.5 to 58.8 its (NDDB report)

Women's cooperative societies have made women's participation in different socio-economic activities more challenging. Cooperatives are considered as an important tool for effecting sustainable rural development by utilizing locally available resources. Their role is instrumental in getting higher income, ensuring equitable distribution of income and employment opportunities. Increasing women involvement and motivation in cooperative can bring remarkable changes in their socio-economic status and assist them in equity, social justice, self-help development and encouraging the women's organizations with requisite technical and management skill and credit facilities to their activities and operations.

With the growth of cooperative institutions, Agricultural Cooperatives Women's Association (ACWA) was set up in Japan after World War II to improve the socio-economic conditions of women. Ahamdnagar Cooperative Bank of India is the model example of the exclusive cooperative by women and for women. The Bank was established in 1974 by a housewife with 100% women staff.

In India and Indonesia, special women cooperative societies are the sources of inspiration of women welfare. Since 1988 the National Dairy Development Board (NDDB) through its cooperative development programme in different parts of India, decided to involve women dairy farmers through cooperative movement.

Sangwan *et al.* (1986) observed that 90 % of rural women in Haryana participated in dairy husbandry activities and livestock care.

Sen and Rani (1990) found that women's involvement in dairying and animal husbandry activities was intensive and if these activities women do together and 93 per cent of workers were women.

World Bank (1991) reported that women accounted for 93 % of total employment in dairy production.

Rangnekar *et al.* (1991) in their study revealed that in livestock rearing, disposal of milk and milk products and feeding of animals are decided by women, whereas sale and purchase of the animals, the decision varied between region and group. They reported that 21 per cent of decision of dairy farming activities were taken by the women as compared to men. The important independent women decisions are cattle/shed management (30.2 per cent); sanitary management (29.2 per cent), feeding of animals (25 per cent) pre and postnatal management of animals 20.8 per cent.

In another study Birader *et al.* (1986) pointed out that rural women having been highly employed in dairy farming activities devote 90 % their time on cattle care making it more or less a female domain.

V.M.Rao (2003) revealed in his study on WDCS in Rajasthan that as a result of the efforts of the unions and the government, 1163 WDCSs were organized by 2001-02. Women membership stood at about 80,000 constituting 20 % of the total membership. Further he pointed out in his study on impact of WDCS of Rajasthan on the members that impact is measured in terms of ownership of animals consumption pattern, change in workload, participationin society is activities various benefits availed of, interactives with outsiders, etc. The average age of the respondents is the highest in Bharatpur and the lowest in Jaipur with on aggregate literacy rate of 41 %. Average size of holding, at the state level, is worked out as 5.76 acres. Income from dairying comprised a substantial percentage of total income. There has been improvement in animal holding, production of milk, consumption and prices realized during post-WDP.

Further, the same study revealed that women spend 606 minutes/ day in pre-WDP which increased to 798 minutes/day in post-WDP. Women took interest in the activities of WDCS and 79 % of them attended at least one annual general body meeting during the lost five years. About 46 % of the women reported receiving payment from WDCS. The money received from the WDCS was spent on purchase of feed or fodder for animals, household provisions, payment of electricity bills, etc women are happy with these benefits which is reflected in higher intake of quality food, improved health conditions, better clothing, enhanced savings and finally leading to increased status.

Usha and Singh (1982) reported that the farm women participation rate in dairy enterprises was as high as 65.49 per cent as against 31 and 3.51 per cent only for men and children respectively. This means that the dairy enterprises could generate gainful employment for the rural women and there by increase the income of the families.

Khedre and Pal (1992) on a study of role of farm women in dairy development revealed that there is a challenging task before the extension agencies, particularly those who are engaged in dairy development, to improve the knowledge and skills of farm women and to bring about changes in their attitudes favourable towards modern dairy innovations. These changes in the behavioural complex of farm women coupled with supplies of needed inputs including dairy credit and services will go long way in adoption of improved dairy farming technologies. They reported that by increasing the employment and productivity of their milch animals their income could be raised to fulfil their day to day needs for a better living standard.

Kumar *et al.* (1999) found that majority of the beneficiaries of WDCS had high level of knowledge about village dairy cooperative and its sphere of activities.

Chinnadurai *et al.* (2004) pointed out that the socio-economic variables and mass media exposure showed positive and significant relationship with knowledge level of women involved in dairy farming.

Sah *et al.* (1999) reported the impact of scientific dairy farming training on 40 women of KVK, NDRI. Study showed that knowledge gain and retention were high and attitude level of symbolic adoption was increased significantly due to training.

Bhatt (1975) told that maintenance of milch animals, their production and sale of milk and milk products were solely the responsibility of housewives. A survey conducted in 1969-70 in Kaira district of Gujrat showed that women had derived considerable employment opportunities and income out of A.H. enterprises.

Kumari *et al.* (1988) observed that 80 % of all important dairying operation in dairy milk production was performed by women.

The study of Sudharani *et al.* (2000) revealed that empowerment of women through dairy farming in rural area was very much significant.

Chinadurai *et al.* (2004) observed that socio-economic variables showed positive and significant relationship with knowledge level of women involved in dairy farming.

Tripathi (2001) conducted a study on information receiving behaviour of Rural Women Influencing Productivity of milch animals. She pointed out that a highly significant relationship was found between utilization of mass media sources and milk productivity irrespective of MC and NMC systems. Thus mass media sources should find a paramount place in enriching scientific knowledge about dairy farm technologies for further augmenting productivity of milch animals. She also revealed that less use of institutionalized sources by rural women may be due to her inability to spare working hours to seek new knowledge, non-accessibility of these sources in addition to these lack of sources with them to match the new knowledge. Thus, there is need to strengthen dairy farming extension services and training sources for field functionaries both in MC and NMC areas.

Gosain and Singh (2002) reported that rural women in India handle a large quantity of milk every day. Having short shelf life, the milk is often converted into various dairy products using the traditional methods of milk processing. They conducted the study with an aim to ascertain the various methodologies prevalent in the rural areas, the training needs of the rural women and also to test selected dairy processing technologies. The technologies for preparation of different dairy products were demonstrated to 150 rural women. It was found that panner and channa murki were liked extremely. The present study suggests that training programmes for different dairy products need to be extensively organized among the rural women in the country duration in the form of different dairy products.

Nanavaty (2002) reported in her study on "women's leadership in dairy sector" that veterinary care is an important need of the cooperative members. As the vets are a limited number so there is a need to train women as paravets. These women would be locally available and also this would be an additional income for women by skill up-gradation. She also emphasized that a need to adopt a strategy of decentralization to combat privatization and globalization. Also small scale chilling plants and processing plants are to be promoted in cooperative managed. This will lead to massive growth of trade and business through the cooperative network.

Om Prakash (1988) stated that there is need for training of farm women in scientific dairy farming practices in Karnal district of Haryana and reported that farm women possessed a low level of knowledge in the area of animal health care, whereas maximum knowledge was found in management practices on the basis of a total sample under study. He further reported that farm women perceived maximum i.e. 78.57 % training needs in animal health care practices, whereas, it was minimum i.e. 54.43 % with regard to feeding practices. However, the overall training needs in respect of scientific dairy farming practices as perceived by farm women was reported to an extent of 63.78 per cent. This implies that the farm women should be provided more skill - oriented, need-based training for higher adoption rate of scientific dairy farming practices to increase milk production for higher returns at their ends.

Dash (2000) reported that management of livestock is one of the important allied agricultural activities in which Indian Women contribute a lot i.e. entire management of dairy animals, starting from cutting, collection and chaffing of fodder to feeding milking, preparation of milk products, cleaning of cattle shed and cattle, collection of dung and dung cake etc.

Chinadsurai *et al.* (2002) in their studies on that "Role of women in dairy farming" observed that farm women having dairy farm with a herd size of minimum three milch animals and registered as members of dairy cooperative were successful. Farm women participated in different operations namely feeding, milking, preparation of milk products, cleaning of dairy sheds, utensils and dairy animals. The constraints expressed by women were low price of milk, non-availability of green fodders in summer, lack of transport facilities, lack of financial resources and non availability of veterinary health care services.

The Dairy Federation in many states in India like the Odisha State Co-operative Milk Producer's Federation (OMFED), the Pradesh Cooperative Dairy Federation (PCDF) in Uttar Pradesh and the Bihar State Cooperative Milk Producers Federation Ltd. (COMPFED) have taken the initiative for the formation of Women's Dairy Cooperative Societies (WDCS) wholly controlled and managed by women.

Relevance of the Study

Odisha is the 10th largest state in India has an area of 15.57 million ha and human population of 36.80 million in 2001. Despite of having bounty of natural resources there is high incidence of poverty, malnutrition, starvation and death. Half of its rural population live below the poverty line. Agriculture along with animal husbandry is the most important income generating activity in rural Odisha. It is reported that 80 per cent of all rural households own livestock and earn income out of them.

Production in the livestock sector in Odisha takes place in millions of scattered holdings (3.68 million hcts.) spread throughout the length and breadth of the state, most of them landless, marginal or small farmers. The average holding size is 3-4 animals of different age groups comprising of cattle or buffalo or both. Milch animals constitute an integral component of the livelihood mechanism of the rural poor and over 66 per cent of the milch animals in the state owned by landless, marginal and small farmers. There are very few large cattle/buffalo holdings in the state.

In Odisha, dairying has been a part of state's anti-poverty programme and it has been accepted as an instrument for improving socio-economic conditions of the rural poor farmers. Various schemes have been implements by the State Deptt. of Veterinary Services and Animal Husbandry and other agencies like ISPO, OMFED for the development of dairy sector. These schemes have mainly concentrated their efforts on providing technical and other inputs to the dairy farmers, but the marketing of milk is exclusively being taken care by OMFED, established as a part of Operation Flood Programme of the Government of India. The OMFED undertakes marketing of milk as well as providing technical input facilities like health care, supply of balanced cattle feed, fodder kits, and provide technical guidance as and when required by them.

Despite of having various programmes implemented in Odisha for dairy development, the milk production in Odisha in 1995-96 was 6.245 lmt which is less than 1 per cent (0.94 per cent) of the total national production, making Odisha 17th rank in milk production in India. The contribution of the different species/types of animals to total milk production in the state in 1995-96 was local cattle 2.96 lmt (47 %), cross bred cattle 2.04 lmt (33 %), buffalo 1.22 (20%) and goat 0.025 lmt (0 %). Milk production in Odisha grew from 0.32 mmt in 1980-81 to some 0.65 mmt in 1996-97, over 100 per cent at a simple average annual growth rate of 5.08 per cent per annum, even though annual growth rates from one year to the next, fluctuate violently. The total milk production in 2009-10 was about 16.51 lmt, a per capita availability was only 109 gm per day against the nutritional requirement of 285 gm per cap per day. As per ICMR recommendation Odisha, however, consumes imported milk solids, mostly powders and infant foods raising the per capita consumption to 109 gm per day.

Reasons attributed to the low productivity of the dairy sector in the state include shortage of feed supply, low production potential cows, cumulative decline of grazing land, rise in population and the dominance of perennial and plantation crops in the cropping pattern. Besides the above factors, less interest of the people in dairying, low quality of care of dairy animals and less support and patronage of Govt. might also be the factors for such low productivity.

In Odisha, keeping cow in home has been accepted as a sacred culture and use of cow dung for washing house floor in every day morning is a usual practice of rural women. Women are more involved in taking care of cows/animals in home. They are unrecognized managers who entirely look after cow's health, stay, feeding, watering, milking and selling milk and bring income to home. A gender study conducted by "the gender study team, AHD, Odisha (Sept. 1999) found that all the day to day activities related to caring, feeding, cleaning, health and production of livestock are predominately performed by women. Clubbing activities performed on a routine basis that impinge on health and productivity of the animal(cleaning of shed, cleaning animal, preparation of cow-dung cake, letting animal in and out of shed, milking, feeding, watering, grazing, collecting green fodder, detection of sick animals, care of sick animal, care of special category of animals like calf etc.). One finds that in 60% of households women perform the activities, and in 30% of the households men perform the activities, and in the remaining 10% of households both share the activities. Moreover, women have greater practical knowledge about the needs, problems and requirements of their livestock because of their day to day involvement with all the activities that are crucial for the survival, health and productivity of the livestock.

With such rich experience of women in animal husbandry practices and their satisfactory involvement as reported by various studies, NDDB since 1988 started promoting the dairy development among women dairy farmers through the cooperative movement.

In Odisha, the Odisha State Cooperative Milk Producers Federation was incorporated under the State Cooperative Societies Act in 1980, as a part of the Operation Flood II Project for Odisha. The milk cooperative setup under operation flood has a three tire setup with the federation as the State level apex body, the district unions and the village dairy cooperative societies are at the grass root level. The federation has now eight milk Unions affiliated to it and some 986 functional village dairy cooperative societies, out of which 330 are women dairy cooperative societies. Total male members of all the societies are 92, 476 of which 20,350 are women(22%).

Besides that, a project called Odisha Women's Dairy Project sanctioned by the Department of Women and Child Development, under the Ministry of Human Resources Development, Govt. of India, New Delhi, is implemented by Odisha State Cooperative Milk Producers' Federation Ltd.through District Cooperative Milk Producers (OMFED) Union Ltd. The project first started in the undivided district of Sambalpur in the year 1995-96 and subsequently implemented in the undivided districts of Cuttack, Puri, Dhenkanal, Keonjhar and Ganjam in the year 1996-97.

The main objective of the project is to improve the socio-economic status of rural women through efficient and modern dairy management practices by formation of Anand pattern all women cooperative societies supported with training and skill upgradation programmes. As per the progress report of OWDP as on June 2001, there were 624 WDCS out of which 582 nos. of WDCS were functional and 554 WDCS were making profit.

From the above preliminary informations it is evident that WDCSs in Odisha are making a good strives for empowering women through dairying. Their progressive success might have the impact in improving the socio-economic status of the members. Till yet, no indepth studies have been conducted in Odisha to know the impact of the WDCSs on their members in term of socio-economic living. Neither the studies have been made to know the performance of all these women dairy cooperative societies with respect to their organizational pattern, management, economic, sustainability, extent of empowerment to women, socio-economic upliftment etc., nor any step has been taken to come out through research with a guideline for effective formation of women dairy cooperative society. With this background, an attempt is made to undertake a research study on "PERFORMANCE OF WOMEN DAIRY COOPERATIVE SOCIETIES IN ODISHA- A CRITICAL ANALYSIS", which would address the following issues;

1. How the WDCSs are being organized and functioning at the village level to safeguard the interest of the cooperative members and the public ?
2. What are the nature and extent of services being provided by the WDCSs to the dairy farmers?
3. What are the nature and extent of support extended by the implementing agencies for promotion of WDCSs?
4. To what extent the management of WDCSs are able to fulfill the expectations of the cooperative members?
5. What are the level of performance of these WDCSs in different aspects like marketing, profit making, transportation etc. ?
6. What are the problems of dairy women farmers and how these problems were being solved by the WDCSs?
7. What are the causes of increasing the number of WDCSs more and more and what are their future sustainability?
8. What are the indicators to measure the success or failure of the WDCSs?

9. What should be the management strategies of WDCSs to face the challenges and opportunities created by changing socio-economic and socio-political environments under the liberalized Government policies?
10. What are the actual possibilities for the members to be able to change their socio-economic and political life through the institutional structures and impact of WDCSs in capacity building with respect to decision making, developing leadership ability, information collection and sharing etc.

The cooperative sector in general and the dairy cooperative in particular continues to occupy an important place in India's economy and rural development strategies. In the context of economic reforms of liberalization and globalization, the cooperatives at the rural locations are to be reoriented in order to meet the challenges of competition, profit maximization, sickness, cumulative loss and collapse. There is an immediate need for technological interventions in the dairy cooperatives at different levels like;

1. Organization Development Intervention (ODI) at the stage of formation of cooperatives which is required to develop staff or member morale and motivation and to unleash the energy and enthusiasm of the members to check organizational rot and create conditions congruous for their working.
2. Information Technology Intervention (ITI) at the functioning stage is one of the important factors for updating the cooperatives with the surrounding development. Computerization of the dairy cooperatives is one of the examples of technology intervention. The amazing speed and accuracy with which the computers are able to process, retrieve, transfer and reproduce data, have made them highly instrumental in augmenting the efficiency of the cooperatives.
3. Modernization of cooperatives with the intervention of new technologies with respect to balanced cattle feed formulation, modern equipments like milking machines, transportation, preservation, processing of milk, equipments for preparation of milk products, modern technology of dairy farm management, feeding, breading and disease control etc,. can be taken up in order to make the society economically viable and sustainable.

In Odisha, as it is understood, the WDCSs are functioning in the same traditional way as before. No studies, till yet, has been conducted to bring the facts into light about various organizational and operational aspects of WDCSs. Even in this information revolution day or IT day, the dairy cooperative societies have failed to access different information through mass media or internet. The functioning of the WDCSs is very slow, sluggish and most of the societies are suffering from organizational sickness.

Therefore, the present study is proposed to make an indepth investigation on the performance of women dairy cooperative societies in general with the following objectives in particular;

Objectives

1. Socio-economic profile of the members of WDCS involved in dairy production activities.
2. Gap in process of procedures, prescribed or followed in formation and functioning of WDCS.
3. Identification of key performance areas in the WDCS.
4. Determination of managemental practices and milk production profiles of WDCS members.
5. Extent of participation of women members in dairy activities.
6. Entrepreneurial characteristics of Women Dairy Cooperative Society members.
7. Impact of WDCS on socio-economic aspects of the members.
8. Correlating variables and their interrelationship in predicting smooth functioning of WDCS.
9. Constraints faced by the WDCS members.

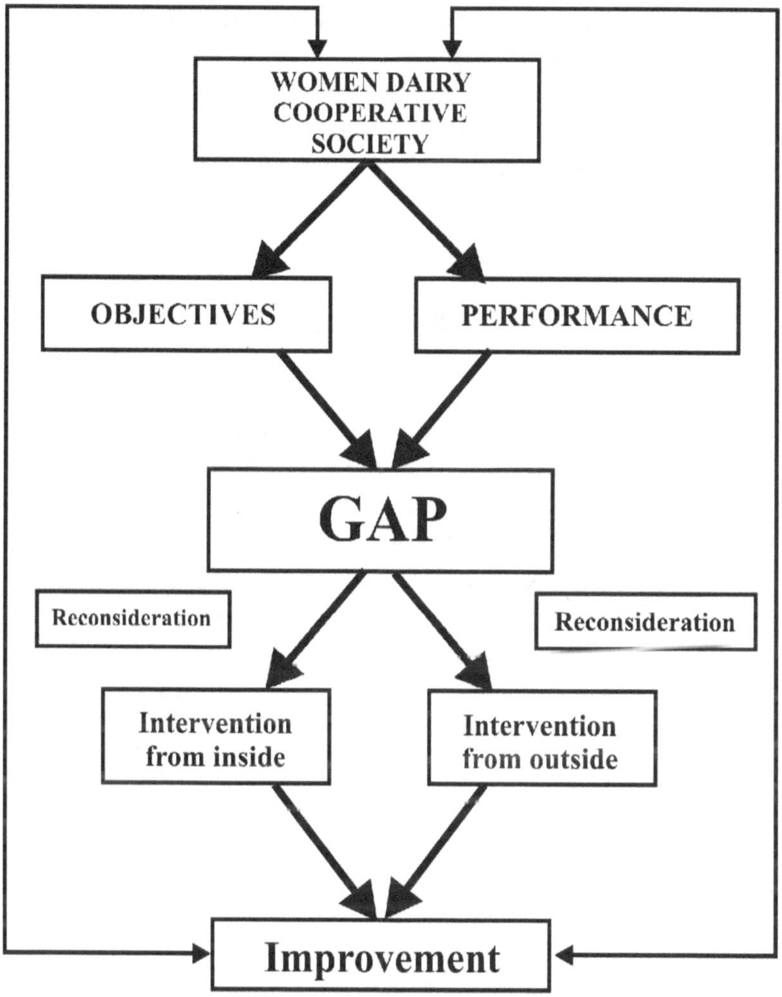

Fig. 2.1 : Conceptual frame work of the study

Practical Utility of the Study

The results of the study will provide an insight on over all functioning of the WDCSs which will help to the implementors/ authorities to understand and take appropriate decision for better output of the societies. The results of the study will develop confidence among women entrepreneurs to take cooperative ventures for generating more income. The study could be used as a reference manual for venturing to establish WDCSs in future. Further, the results of the study will be of great help to understand how cooperative action through small contributions can bring about a great change in socio-economic life of the people. The results of the study will be helpful to the Govt./NGOs to draw conclusion that cooperative efforts of women in dairying could empower the women and help them to develop and be equal partners as to their male counter parts. The results of the study will provide some constructive recommendations to the concerned authorities for bringing about structural and functional change in the WDCSs for better output.

Though the study has great significance to all the concerned individuals and agencies involved in development of dairy and rural development sectors as well academically relevant for the interest academicians and planners and policy makers, it may also have some limitations and shortcomings. These are being detailed below:

1. The present study may have a limitation of time, money, accessibility and other resources that might affect the quantity of coverage of data. However, every possible step has been taken to maintain the quality of data intact.
2. As a social science research, the findings of the research study were based on the data collected from the respondents of the study and the records made available to the researcher. The prejudices and biases of the respondents and the reliability of the data collected from the secondary sources are not ruled out.
3. The results of the study might not be applicable to a wider geographical area because the study was conducted in two different agro-climatic conditions. However, the findings in general will give an insight of functioning of WDCSs and will supplement other studies on cooperative societies.

❑❑❑

Chapter - 3

Profile of the Odisha State

The study was undertaken in the state of Odisha. The state lies in between 17°-31′ to 22°-27′ North latitude and 81°-27 to 87°-30′ East longitude. It is bounded by West Bengal in the north east, Jharkhand in the north, Chatisgarh in the west, Andhra Pradesh in the south and Bay of Bengal in the east. The main features of the state is appeared in Table 3.1.

Table 3.1 : Odisha at a Glance (2001 census)

Sl. No.	Facts	Unit	Magnitude
1	Geographical situation	-	Between 17°-31' to 22°-2T north latitude and 81°-27' to 87"-30' east longitude
2	Geographical area	Sq.km	1,55,707
3	Population (2001 census)	Number	3,67,06,920
4	Male population	-do-	1,86,12,340(50.71 %)
5	Female population	-do-	1,80,94,580 (49.29%) . .
6	Density of population	Person/sq.km	236
7	Literacy rate	Percentage	63.61
8	Male literacy rate	-do-	75.95
9	Female literacy rate	-do-	50.97
10	Sex ratio	Female/1000 males	972

Source : Economic Survey : Directorate of Economics and statistics, Odisha-2001).

Physiography of the State

Odisha is an extensive plateau which slopes gently into coastal plain along the Bay of Bengal. The river Mahanadi is flowing west to east through the plateau divide the state into two definite parts. The northern part is an extension of Chhotnagpur plateau and the southern part is a part of Eastern Ghat region.

The state has been divided into ten agro-climatic zones. The distinctive features of various zones are as follows.

Fig. 3 : India Map showing Odisha State.

i) North western plateau

The zone comprises of agricultural districts of Sundargarh, Beneigarh and Panposh in Sundargarh revenue district and Kuchinda under Sambalpur district. The climate is hot, moist and sub-humid having mean annual rainfall of 1600 mm, mean maximum temperature of 38°C and mean minimum temperature of 15°C. The soil is light textured with red and yellow soils. Major crops grown are rice, toria, wheat and vegetables.

ii) North central plateau

It consists of agricultural districts Rairangapur, Baripada, Karanjia in Mayurbhanj district and Keonjhar, Champua in Keonjhar district. The climate of the zone is hot, moist and sub-humid. The mean annual rainfall is 1534 mm with mean maximum and mean minimum temperature of 36.6°C and 11.1°C respectively. Red loam and light textured soils mostly found in the zone. Rice, niger, maize, horse gram, black gram, arhar, sesamum and vegetables are the major crops grown.

iii) North eastern coastal plain

Agricultural district Anandapur under Keonjhar district, Balasore and Bhadrak districts are coming under this zone. The climate of the zone is moist and sub-humid with mean annual rainfall of 1568 mm. The mean maximum temperature is 36°C and mean minimum temperature 14.8°C. Alluvial soils and narrow strip of sedimentary sandy loam to clay textured soils are the characteristics of this zone. Major crops grow are rice, jute, groundnut, toria, vegetables, blackgram, greengram and betelvine.

iv) East and south eastern coastal plain

Cuttack and Banki agricultural districts in Cuttack districts, Jagatsinghpur, Kendrapara, Puri districts and Berhampur, Chhatrapur agricultural districts in Ganjam district comprises this zone. The climate is hot and humid with mean annual rainfall 1577 mm and mean maximum as well as mean minimum temperature of 39°C and 11.5°C respectively. The soils of the zone are mainly loam, clay loam and coastal alluvial with a narrow strip of saline inundated areas. Rice, greengram, blackgram, groundnut, jute, sesamum and vegetables are the major crops grown in this zone.

v) North eastern ghat

The zone consists of Khurda, Nayagarh, Kandhamal, Boudh, Gajapati districts, Rayagada and Gunupur agricultural districts in Rayagada district and Aska agricultural district in Ganjam district. The climate of the zone is hot, moist and sub-humid with mean annual rainfall of 1597 mm. The mean maximum and minimum temperature of the zone is 37°C and 10.4°C respectively. The soils of the zone are brown forest soil, sandy loam, loamy and clay with medium texture. Rice, ragi blackgram, greengram, til, groundnut, sugarcane, toria, turmeric and vegetables are the major crops grown in this zone.

vi) Eastern ghat high land

Koraput and Nawarangpur district except Dabugaon area are under this zone. The climate is warm and humid with mean annual rainfall of 1521.8 mm. The mean maximum and minimum temperature are 34.1°C and 7.5°C respectively. The zone consists mostly red soils with low organic matter content. Major crops like ragi, rice, small millets, vegetables, maize, arhar, niger, toria and sweet potato grown in this zone.

vii) South eastern ghat

The agricultural district Jeypore in Koraput district and Malkangiri district are included in this zone. The zone is having warm and humid climate. The mean annual rainfall, maximum and minimum temperature are 1710.4mm, 34.1°C and 13.2°C respectively. Red and red yellow mixed soils are the characteristics of this zone. Rice, arhar, niger, toria, mazie, sweet potato and vegetables are the major crops grown.

viii) Western undulating zone

The zone comprises of the agricultural districts Bhawanipatna, Dharmagarh in Kalahandi district, Padmapur in Bolangir district, Khariar in Nawapara district and Dabugaon area of Nawarangapur district. Hot, moist and sub-humid climate with mean annual rainfall of 1352.3 mm is the main features of the zone. The mean maximum and mean minimum temperature are 37.8°C and 11.9°C respectively. Black, heavy light soils and mixed red black soil with medium to heavy texture are found in this zone. Rice, ragi, arhar, bengalgram, toria, onion, cotton, vegetables and groundnut are the major crops grown in this zone.

ix) Western central table land

Sambalpur, Baragarh, Bolangir and Sonepur districts belong to this zone. The climate is hot, moist and sub-humid with mean maximum temperature of 40°C and mean minimum temperature 12.4°C. Red, heavy textured calcareous and light to medium textured soils are mostly seem in this zone. Average annual rainfall is 1282 mm. Major crops grown are rice, til, blackgram, wheat, greengram, groundnut, sugarcane and vegetables.

x) Mid-central table land

The zone consists of agricultural districts of Dhenkanal in Dhenkanal district, Angul and Kamakshyanagar in Angul district, Banki and Athagarh in Cuttack district and Sukinda block in Jajpur district. The climate of the zone is hot, dry and sub-humid. The mean annual rainfall is 1421 mm with mean maximum and mean minimum temperature of 38.7°C and 14°C respectively. Light textured, lateritic, medium textured, red loam, mixed red and black soils are found in this zone. Rice, groundnut, sesamum, black gram and vegetables are the major crops grown in the zone.

Livestock Resources of the State

Development of animal resources has considerable relevance and significance for development of an agrarian economy. Next to crop production, animal husbandry is the most important income generating activity in rural households. Over 80 % of the rural households in Odisha keep livestock of one species or another. The State Agricultural Policy, 1996 aims at increasing the production and productivity in dairy, poultry, and other animal husbandry activities through scientific breeding, balanced feeding, adequate health care, better marketing facilities, and improved management practices. The objective of the Animal Resource Development Programme is not only to increase the animal resources and to increase availability of their products but also to provide employment and income generating opportunities through both main and supplementary enterprises.

During the 9th Plan period, emphasis was laid on bringing about genetic improvements in the livestock breeds and to enhance their productivity. Besides, efforts have also continued to upgrade skill of farmers by imparting required training. During Tenth Plan the following strategies have been envisaged.

Strategies taken during Tenth Plan Period

i) Expanding and strengthening of cross-breeding programme through Frozen Semen Technology and improvement of local cattle and buffalo wealth of the State for production of high-yielding milch cows and working bullocks.

ii) Control of major livestock diseases through systematic research and monitoring.

iii) Strengthening of disease reporting network.

iv) Revamping of production infrastructure for biological with a view to enhancing quantum of production of life-saving vaccines for livestock.

v) Genetic improvement of small animals for enhancement of meat production.

vi) Introduction of state implementing agency "Odisha Livestock Resources Development Society" for facilitating implementation of project Cattle and Buffalo Development.

vii) Utilization of surplus milk by establishment of Milk Powder Plant.

viii) Reduction of livestock mortality and development of nutritional status of tribal through cluster approach of integrated animal husbandry in KBK districts.

ix) Professional efficiency development by enforcement of the Veterinary Council Act.

x) Creation of resources by introducing user charges for various veterinary services to provide better qualitative and non-interrupted services.

The distribution of livestock population in the State into various species is indicated in Table 3.2 . The livestock population in the state excluding dogs was 230.57 lakh as per the livestock census held in 2007, registering a marginal decrease of about 4.01 % over the livestock census held in 2003. Out of the total livestock population, cattle population was 123.09 lakh (53.38 %), buffaloes population 11.9 lakh (5.16 %), goats 71.27 lakh (30.90 %), sheep18.18 lakh (7.88 %) and pigs 6.12 lakh (2.65 %). Apart from this population, other animals i.e., dogs, horses etc, figures at 9.25 lakh. During 2007, while the cattle and buffaloes population decreased by 13.80 %, 17.30% respectively, the population of goats, sheep and pigs were increased by19.30 % and 3.35 % & 7.55 %

respectively over the previous livestock census held in 2003. Further, it is observed that the percentage of crossbred/improved animals in our state is very poor i.e. out of total cattle population the percentage of cross bred cattle is 13.83 % and in case of buffalo, the percentage of improved buffaloes is 2.3 %. Similarly the percentage of crossbreed goats and sheep were 0.6 % and 0.4 % respectively and in all species the percentage of female animal population is much higher than the male.

Table 3.2 : Livestock population in Odisha (Figures in lakhs)

Census year	Cattle		Buffaloes	Goats	Sheep	Pigs	Total
	Total	Cross bred					
1	2	3	4	5	6	7	8
1982	129.3	N.A.	13.33	49.31	19.9	4.1	215.94
1991	135.77	5.63	15.09	48.08	18.41	5.9	223.21
1995	147.66	7.44	16.52	54.12	18.65	5.72	242.67
2001	140.03	8.71	13.88	58.8	17.79	6.02	236.52
2003	142.81	10.43	14.39	59.74	17.59	5.69	240.22
2007	123.09	17.03	11.9	71.27	18.18	6.12	230.57

Source : Directorate of Animal Husbandry and Veterinary Services, Odisha.

Availability of Animal Products

The production and per capita availability of milk, meat and eggs in Odisha for the period from 1998-99 to 2009-10. has by and large gone up steadily (Table 3.3). The per capita availability of milk has exceeded 109 gms/per head/per day while that of egg is 57 eggs/per head/per annum during 2009-10. During 2009-10, the per capital availabilityof milk and egg at the National level was 263 gms/day and 74 nos./annum respectively while it was 109 gms/day and 57 nos/annum in our state. While the production of milk and egg was 112.00 million tonne and 59,000 million numbers at all India level, in our state it was only 1.65 million tonne and 2319.32 million numbers respectively during 2009-10.

Table 3.3 : Year-wise production and per capital availability of milk, meat and eggs in Odisha

Year	Productionof milk (thousand MT)	Per capita Availability of Milk(gms/day)	Productionof meat (thousand MT)	Per capita Availabilityof meat (Kg./ Annum)	Production of eggs (in millions)	Per capita Availability of eggs (nos/ annum)
1	2	3	4	5	6	7
1999-00.	847.78	65	36.67	1.029	648.31	18
2000-01	875.13	67	38.38	1.067	760.10	20
2001-02	928.77	70	42.64	1.176	858.02	24
2002-03	941.00	71	44.72	1.224	909.49	25
2003-04	995.08	74	44.89	1.218	931.06	25
2004-05	1282.76	94	50.53	1.358	1251.96	33
2005-06	1342.30	98	52.04	1.385	1278,71	34
2006-07	1424.64	103	55.01	1.450	1424.64	38
2007-08	1620.44	115	58.82	1.530	1549.48	29
2008-09	1598.05	113	115.00*	2.933	1993.26*	51
2009-10	1651.33	109;	128.00	3.170	2319.32**	57
2010- 11(P)	1670.00	109	138.00	3.290	2357.10	56

Source : Directorate of Animal Husbandry and Veterinary Services, Odisha.

* Include poultry meat. ** Include private sector production.

Animal Health Care

Provision of adequate health care is a prerequisite for increasing the production potentiality of livestock. The spread of veterinary services in the state has resulted in the control of some communicable diseases and improvement in animal health. By the end of 2010.11, the state had built up a network of 540 Veterinary Hospitals/Dispensaries with 613 Veterinary Doctors, 2,939 Livestock Aid Centres (LACs) with 2,566 Livestock Inspectors and Technicians, 3 Clinical Investigation Laboratories, 3 Disease Research Centres, one Animal Disease Research Institute at Phulnakhara in Cuttack district and two Biological Product Institutes (one at Bhubaneswar and the other at Berhampur) to provide curative and preventive health services. In order to provide better health care to the animal population, there is need to open at least one Livestock Aid Centre in every Gram Panchayat in the State in a phased manner. During 2010.11, preventive inoculation was given to 302.14 lakh animals,5.00 lakh castrations were performed and 53.03 lakh animals and birds were treated.

Table 3.4 : Animal Health Care

Item	2004-05	2005-06	2006-07	2007-08	2008-09	2009-10	2010-11
1	2	3	4	5	6	7	
Treatment of animals (in lakh nos)	49.66	47.88	47.81	51.29	55.48	55.69	53.03
Vaccination production (in lakh doses)	105.55	92.17	136.73	184.65	193.78	210.11	235.27
Vaccination done (in lakh no.)	114.02	115.98	148.18	202.35	242.70	276.34	302.14
Infertility camp organized	4587	3785	4197	3229	3508	3118	3172
De-worming of animals (in lakh nos)	2.13	2.30	2.49	3.09	7.28	8.84	7.24
Animal castrated (in lakh nos)	4.66	4.74	4.55	4.53	4.41	4.92	5.20

During 2010-11 7.24 lakh animals have been dewormed and 3172 infertility camps have been organized. Odisha Biological Product Institute, Bhubaneswar and Berhampur are functioning in the state which are producing lifesaving biological vaccines and medicines against contagious diseases have been produced 235.27 lakh doses of vaccine of different specification during 2010-11, and an amount of Rs.8.58 lakh has been provided in the Annual Plan 2004-05. Rinderpest Eradication Scheme, a 100 % Central Funding scheme, is in operation in the State with the aim to eradicate rinderpest diseases in cattle and buffaloes including small animals by arousing mass awakening among the farmers in the state.

Cattle and Buffalo Development

According to the livestock census held in 2007, the cattle and buffalo population together was 134.99 lakh which constituted 58.54 % of the total livestock population of the State (Table 3.5). Out of this, the female animals were 64.41 lakh including 29.91 lakh milched and 35.07 lakh were dry. Besides, out of the total cattle population in the State,17.03 lakh (13.83 %) were crossbred/improved variety, of which 8.25 lakh were female and 3.32 lakh were milched. This indicates that, out of total female cattle and buffaloes, 12.80 % belonged to cross bred/improved variety and 5.15 % were milched.

Table 3.5 : Cattle and buffalo population as per 2007 census (No. in millions)

Category	All Total		Female		Milch		Breedable	
	Total	C.B	Total	C.B	Total	C.B	Total	C.B
1	2	3	4	5	6	7	8	9
Cattle	123.09	17.03	58.53	8.25	27.1	3.32	31.72	4.47
Buffalo	11.9	NA	5.88	NA	2.81	NA	3.35	NA
Total	134.99	NA	64.41	NA	29.91	NA	35.07	NA

The productivity of the cattle of our state is very low. In order to upgrade the production potential of cattle and buffaloes, two schemes, namely, Artificial Insemination through Frozen Semen Technology and Strengthening of State Cattle Breeding Farms are being implemented. While the former aims at production of superior breeds of female progeny, the latter is intended to provide breeding bulls.

Besides the Veterinary Hospitals/Dispensaries, there are 252 pure artificial insemination centres and 2,180 LACs in the State have the facility of artificial insemination through frozen semen technology.

During 2003-04, 4.37 lakh artificial insemination and 6883 natural services were performed against the target of 8.06 lakh artificial insemination. Against the above insemination 1,86,375 progeny were born. It has been targeted to conduct 8.3 lakh artificial inseminations during the year 2004-05.

During 2010-11, 8 State Livestock Breeding Farms were functioning in the state under the Government sector to undertake the activities of Frozen Semen Insemination. Different breeds of cattle and buffaloes are reared in these farms to produce genetically superior bull calves. During 2003-04, 57 bull calves have been produced for natural insemination in remote and in accessible areas of the State.

Two externally aided projects namely, Indo-Swiss Natural Resources Management Programme on Animal Husbandry Development with financial assistance from the Government of Switzerland and the Integrated Livestock Development Project with financial assistance from Government of Denmark are in operation in the State. An amount of Rs.81.92 lakh has been proposed during 10th Plan for these projects with DANIDA assistance.

The Odisha State Corporative Milk Producers Federation Ltd. (OMFED) came into being on 01.01.1980 with objectives of promoting production, procurement, processing and marketing of milk and milk products for the economic development of the rural farming community. The Federation had received funds from the Indian Dairy Corporation/ National Dairy Development Board (NDDB) for implementation of Operation Flood Programme in undivided districts of Cuttack, Dhenkanal, Keonjhar, Puri and Sambalpur. This Federation also received funds towards Co-operative Development Programme from NDDB which has been successfully implemented in the undivided districts of Cuttack, Puri, Sambalpur and Keonjhar for implementation of Technology Mission on Dairy Development (TMDD) in the Operation Flood districts. Apart from NDDB, the Federation is also receiving financial assistance from the Government of Odisha as Grant-in-aid and share capital for different activities. At present it is functioning with 9 dairy plants with a capacity of 225.00 thousand litre per day, 25 milk chilling plants with capacity of 115.00 thousand litre per day and 13 bulk coolers with a capacity of 11.50 thousand litre per day. The dairy plant at Bhubaneswar is under expansion from the present capacity of 75.00 thousand litres per day to 125.00 thousand litre per day. Besides, a cattle feed plant with capacity of 100 MT/day is functioning under OMFED.

During 2003-04, 1,097 Milk Producers Cooperative Societies were functioning in the State. These societies procured 1,35,573 kg milk per day on an average. OMFED marketed 1,35,835 liters milk per day on an average. It has also marketed milk products such as ghee, sweet, curd, plain curd, butter, butter milk and paneer etc during 2003-04. Besides the above OMFED has also extended its business by marketing cattle feed and horticultural products like squash, jam, sauce, pickle turmeric powder and mustard seeds. Besides OMFED is also providing health care facilities to the farmers through it's A.I. Centres and sales cattle feed to the farmers in reasonable prices.

Profile of the Districts Under Study

The study was undertaken in the undivided districts of Cuttack and Dhenkanal in the state of Odisha. These two districts are located in two different agro-climatic zones of Odisha. Both the districts differ with respect to different aspects as shown in Table 3.6to 3.9.

Table 3.6 : Administrative setup of undivided Cuttack and Dhenkanal Districts

Sl.No.	Facts	Cuttack	Dhenkanal
1	Geographical area (sq.km.)	11143	10827
2	Sub-division	6	7
3	Tahasils	28	11
4	CD blocks	41	16
5	Grampanchayat	1046	408
6	Municipality	6	2
7	Police station	62	31
8	N.A.Cs	4	4
9	Villages	6556	3125
10	Density	570	210
11	Population	6325069	2206881
12	Sex ratio	972	951
13	Literacy	70.81	69.10

Table 3.7 : Livestock population profile of undivided Cuttack and Dhenkanal Districts

Species	Cuttack	Dhenkanal
Cross Breed (M + F)	247619	54859
Indigenous (M + F)	1513547	1004378
Exotic (M + F)	29391	-
Indigenous (Buffalo)	83285	107798
Improved (Buffalo)	11283	162
Sheep	200007	117694
Goat	770288	441688
Pig	22415	4590
Poultry	1685397	716638

Table 3.8 : Veterinary service profile of undivided Cuttack and Dhenkanal Districts.

Species	Cuttack	Dhenkanal
Veterinary Hospitals and Dispensary	71	34
Veterinary Doctors	99	35
Livestock Aid centres	459	167
Artificial insemination centres	616	178
Livestock Inspectors	381	169
Artificial Insemination done	123041	29881
Progeny born	51772	12084

Table 3.9 : Livestock production profile of undivided Cuttack and Dhenkanal Districts.

Species	Cuttack	Dhenkanal
Milk (Cow) 000 mt	215.31	02.95
Milk (Buffalo) 1000 mt	21.77	17.44
Egg production (No.) in lakhs	737.50	167.52
Meat production in mt	6078	2904

Profile of Selected Milk Unions

Progress of Cuttack Milk Union

Cuttack District Cooperative Milk Producers' Union Limited was registered in 1976 with Cuttack, Jagatsinghpur, Kendrapara, and Jajpur as command area. The union has a total area of 10923 sq.kms. spread over 6108 inhabited villages and comprised of 902 Gram Panchayats. It has breedable cattle population of 11.10 lakhs of which 1.27 lakhs are crossbreds. Further, it has 63000 buffaloes too. The management of the union was transferred to an elected board in 1981.The union has 18 Board of Directors of which 12 are elected. The union has benefited under operation flood programme, DFID, integrated dairy development programme, etc. which helped in expanding installed capacities over the periods. It has 111 employees, 7 chilling centres, (Tirtol, Balikuda, Adaspur, Kendrapara, Banki, Binjharpur and Dasarathpur) and 22 bulk coolers with an installed capacity of 54800 litres/day. The union manufactures full cream milk, toned milk and butter milk and supplies vast majority of its procurement of OMFED. Paradip, Jagatsinghpur and Kendrapara are the marketing centres of the union. The salient features and progress of the milk union during 1991-03 is provided in Tables 3.10. There is a spectacular improvement in all the indicators included during the study period. Registered DCSs increased from 190 in 1990-91 to 419 during 2002-03 and that of functional DCSs from 276 to 485. Similarly, membership increased from 18000 to over 48000 during 1991-03. Women membership constituted only 13 % during 1990-91 which stood at 57 % by 2002-03. Union's daily average procurement was about 19000 litres during 1990-91 which increased to 59365 litres by 2002-03. Out of 13 years for which data are available, union incurred losses during five financial years.

Table 3.10 : Salient features of Cuttack Milk Union

Geographical information	Units (No,)
Total area in sq.km	109229.9
Total inhabited villages	6108
Gram panchayat	902
Blocks	41
Tahsil	26
Police station	51
Livestock statistics upto date	
Breedable Cattle Population	11,10,358
Crossbreed Cattle Population	1,27387
Indegenous cattle population	9,50,729
Buffalo population	62,828
No. of Veterinary Diospensary	71
No. of Livestock Aid Centre	518
No. of A.I. Centres	
A.H. Department	346
Milk Union	116
Total number of Milk Producers' Co-operative Societies Organised	779

Table 3.11 : Districtwise activities of Cuttack Milk Union

SI. No.	Description	Name of the district				Total
		Cuttack	Jagatsinghpur	Kendrapara	Jajpur	
01.	**DCS (Nos)**					
	i. Organised	292	315	114	58	779
	a. MPCS	182	244	63	-	489
	b. WDCS	110	71	51	58(IDDP)	290
	ii. Functional	269	311	105	57	742
	a. MPCS	164	242	62	-	473
	b. WDCS	105	69	43	57CIDDP)	269
02.	Membership (Nos.)	22866	26078	6612	2917	58473
	a. Male	13412	16596	3844	-	33852
	b. Female	9454	9482	2768	2917	24621
03.	Average Milk Procurement (LPD)	36780.0	68831.0	18560.0	3725	12589 6.0

Source: Cuttack Milk Union, Mahanadi Vihar, Nayabarzar, Cuttack.

Progress of Dhenkanal Milk Union

Dhenkanal District Co-operative Milk Producers Limited was established in 1976 and transferred to elected board in the very next year. Operations of the Union cover both Dhenkanal and Angul districts. Like other milk unions, it has 18 Board of Directors of which 12 are elected. The union has 5 chilling centers with an installed capacity 15000 liters/ day. Entire milk procured is supplied to OMFED and the union gets commission of Rs. 1.60/ litre and 50 paise / litre of milk sold in its command area. Thus, union has to meet its expenses through Rs. 2.10/ litre of milk procured and sold.

Table 3.12 : Salient features of Dhenkanal Milk Union

SI. No.	Facts	Cuttack	Dhenkanal
1	Geographical area (sq.km.)	11143	10827
2	Sub-division	6	7
3	Tahasils	28	11
4	CD blocks	41	16
5	Grampanchayat	1046	408
6	Municipality	6	2
7	Police station	62	31
8	N.A.Cs	4	4
9	Villages	6556	3125
10	Density	570	210
11	Population	6325069	2206881
12	Sex ratio	972	951
13	Literacy	70.81	69.10

Table 3.13 : Veterinary service profile of undivided Dhenkanal district

Species	Cuttack	Dhenkanal
Cross Breed (M + F)	247619	54859
Indigenous (M + F)	1513547	1004378
Exotic (M + F)	29391	-
Indigenous (Buffalo)	83285	107798
Improved (Buffalo)	11283	162
Sheep	200007	117694
Goat	770288	441688
Pig	22415	4590
Poultry	1685397	716638

Table 3.14 : Progress of Dhenkanal Milk Union 1991 -04

Sl. No.	Content	1990 -91	1991 -92	1992 -93	1993 -94	1994 -95	1995 -96	1996 -97	1997 -98	1998 -99	1999 - 200	2001 - 01	2001 -02	2002 -03	2003 - 04
1	DCSs registered	67	69	70	70	70	70	71	74	74	74	74	81	87	92
2	DCSs functioning	50	54	55	59	71	82	82	82	91	88	87	87	112	125
3	Membership (Lakhs)	0.05535	0.02864	0.05962	0.06126	0.06542	0.06716	0.07291	0.09860	0.10539	0.10985	0.11806	0.12173	0.12291	0.12451
	Men	-	-	-	-	-	-	0.06069	0.05475	0.05207	0.02407	0.04777	0.04812	0.05325	0.0521
	Women	-	-	-	-	-	-	0.01222	0.04386	0.053 32	0.08578	0.07209	0.07361	0.06966	0.06930
4	Procurement (000Kgs / day)	1.940	1.880	1.900	3.083	3.576	0.658	2.910	2.487	2.307	2.565	2.868	2.264	2.394	2.426
5	Capacity Utlization (0%)	19.40	18.850	19.00	30.83	35.76	26.58	29.10	24.87	23.07	25.65	28.68	22.6 4	24.63	28.19
6	Payment to producures (Rs. Cores./ annum)	0.3528	0.3940	0.5125	0.7795	0.8497	0.6261	0.8889	0.7414	0.6946	0.8559	0.9320	0.7298	0.7428	0.9123
7	Cost of milk (Rs./ Kg)	4.98	5.74	7.39	6.93	6.51	6.45	8.37	8.17	8.25	9.15	8.90	8.83	8.92	8.87
8	DCSs Covered under AI centre	18	16	13	12	13	18	18	43	55	57	52	55	60	67
9	AI done (Lakhs)	0.01436	0.01760	0.01543	0.01326	0.01628	0.2865	0.02968	0.04290	0.05756	0.05505	0.06385	0.05236	0.03127	0.02105
10	Net Profit / Loss (Rs. / Lakhs)	(-) 6.48	(-) 7.65	(-)8.41	(+)0.54	(+)0.65	(+)0.54	(+)0.51	(-)7.20	(-)9.36	(-)17.77	(12.72)	(13.99)	(-)12.92	(-)12.40

❑❑❑

Chapter - 4

Research Methodology

The major focus of the study was to critically analyse the overall performance of Women Dairy Cooperative Societies (WDCS) in two districts of Odisha. The methodology used in carrying out investigation has been outlined in eight sections as mentioned below.

1. Selection of problem
2. Research design
3. Plan of work
4. Locale of the study
5. Sampling procedure
    a) Selection of district milk unions
    b) Selection of milk routes and WDCS
    c) Selection of respondents
6. Tools and Techniques used for data collection
    a) Pilot study
    b) Preparation of interview schedule
    c) Pre-testing of the interview schedule
    d) Collection of data
7. Operationalization and measurement of variables
8. Data processing and analysis using statistical methods

1. Selection of Problem

The important consideration in social research are selection, delineation and conceptualization of problem. A good investigation gives priority on formation of clear-cut, realistic and un ambiguous objectives. Delineation of problem is more important and essential than finding out the solution.

Next to green revolution, India has witnessed a tremendous progress in white revolution during the seventies. The livelihood system of Indian rural families essentially composed of crop and animals. Throughout the country, farm families have given equal even more importance to dairy animals for a variety of reasons. Many families depend on dairy animals for livelihood. So far milk production is concerned, India is at the top in the world. Per capita consumption of milk in India is reported to be 263gm per day.

Production, productivity and profit of the milk need cooperative efforts. This was rightly recognized by Dr. Vergese Kurian who started Ananda Pattern dairy movement in the year 1946. Owing to the influence of Ananda Pattern, many states in India have come out with dairy programmes with varied success. However, it is realized that at different levels even though milk production has been increased, the farmers are not able to get profit as expected. To support the family income, women in India have came forward to add to the profession through women dairy cooperative societies. In India at present about 8089nos of WDCS operating and in that of Odisha 918. In every town of Odisha OMFED milk is available at doorstep for which women dairy farmers have significant contribution. Unfortunately their living standard has not yet been increased as evidenced by 48% of population of Odisha is still live under poverty line. The dairy owners constitute majority of these categories. This has raised doubt in minds of the planners, veterinary scientists and above all common people that what may be the reasons in the WDCS for not enabling dairy owners to come upto their expectations. This question is thought to be more important for the veterinary extension service. Under this backdrop, the problem was selected for detail study as no other such study was undertaken in the state. While selecting the problem, extension capacity building, constraints, issues in WDCS were given due attention for examining the extent of involvement of women in dairy enterprise.

2. Research Design

According to Kerlinger (1963) research design is the plan, structure, and strategy of the investigation conceived so as to obtain answer to the research question and to control variance. The research design of the study was considered in the line of case studies and direct observation methods. Efforts were made to focus the study on the members of WDCS, functionaries and officials involved in the functioning of the societies so that their achievements, experiences, difficulties and suggestions could be collected to make the study realistic and recommendable.

The present study comes within the purview of ex-post-facto research approach. Accordingly, specific objectives were formulated to facilitate the research. In the light of the objectives, the scope of the study was determined, techniques of investigation decided and tools selected for use along with pattern of statistical analysis.

3. Plan of Work

Before proceeding into actual investigation, the efforts were made to collect the relevant materials, related research to gain an in depth understanding on the subject. Attempts were made to have a prior visits to the area under study for formulating the research objectives.

4. Locale of the Study

The study was conducted in the undivided districts of Cuttack and Dhenkanal in the state of Odisha. Odisha, the costal state of India is located between 17°-31′ to 22°-27° north latitude and 81°-27′ to 87°-30′ east longitude surrounded by Bay of Bengal in the east. The state spreads over 30 districts situated in 10 different agro climatic zones. The districts fall under different agro climatic zones are detailed below.

Table 4.1 : Districts under Agro-Climatic zone of the state

Sl.No.	Agro climatic Zones	Districts
1	North western plateau	Sundargarh, Deogarh
2	North central plateau	Keonjhar, Mayurbhanj
3	North eastern coastal plain	Balasore, Bhadrak, Jajpur
4	East & south eastern coastal plain	Puri, Nayagarh, Khurda, Jagatsinghpur, KendraparaCuttack
5	North eastern ghat	Phulbani, Ganjam, GajapatiRaygada
6	Eastern ghat highland	Nawarangpur, Koraput
7	South eastern that	Malkanagiri
8	Western undulating zone	Kalahandi
9	Western central table land	Baragarh, Bolangiri, Sambalpur, Bouth, Jharsuguda
10	Mid central table land	Angul, Dhenkanal

The district of Cuttack and Dhenkanal are located in two different agroclimatic zones of Odisha. Though both the districts have little variation in their geographical area i.e. 11,143 sq.km. and 10,827 sq.km respectively, but there was great difference in possession of number of cross breeds i.e. 2,47,619 in Cuttack district and that of Dhenkanal district is 54,859. In the area of cow and buffalo milk production, a wide difference is existing between two districts . The cow milk production i.e. about 2,15, 000 mt in Cuttack district and 2,000 mt. in Dhenkanal district. However, there is little difference in buffalo milk production i.e. 22,000 mt in Cuttack district and 18,000 mt in Dhenkanal district. The Cuttack district is having 63,25,069 numbers of human population and that of Dhenkanal district is 22,06,881 speaks that the density of population in Cuttack district is more than double i.e. 570 than Dhenkanal district i.e. 210. Although the Women Dairy Project was launched in these districts in the same year i.e. 1996, but the two districts have not progressed equally in the dairy development sector in general and women in dairying in particular. As evidenced from the records that there is a huge gap in the total procurement of milk between two districts after implementation of WDP.

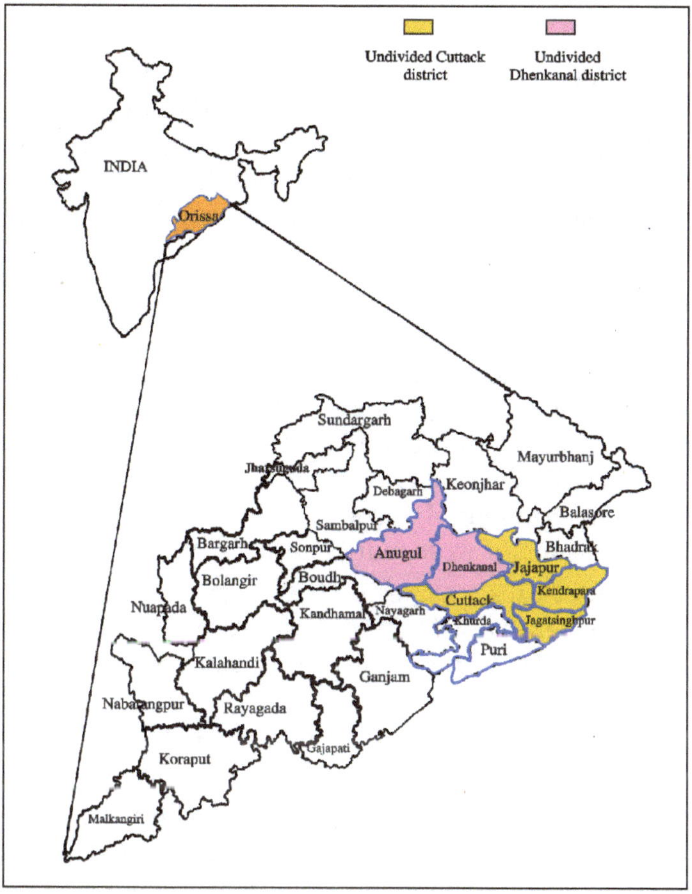

Fig. 4.1 : Odisha Map showing undivided Cuttack & Dhenkanal Districts.

Table 4.2 : Some facts about undivided Cuttack and Dhenkanal districts

Facts	Undivided Cuttack district	Undivided Dhenkanal district
Agro climatic zone	East and South eastern coastal plain	Mid central table land
Geographical area (sq.m.)	11143	108227
Human population	6325069	2206881
Density	570	210
Cross breeds	247619	54859
Milk production	215000 mt	2000mt
Milk procurement	3368000 lt	82000 lt

5. Sampling Procedure

A multistage sampling procedure was followed to select the sample for the study. The district milk unions and the milk routes were selected purposively, whereas the Women Dairy Cooperative Societies and the members were selected randomly from each society. However, the selection of President and Secretary were made purposively from each selected soceity.

a) Selection of district milk unions

In order to ensure guarantee for employment and income generation, minimum wages, welfare and support services, training and up gradation of skill etc., a programme called "SUPPORT TO TRAINING AND EMPLOYMENT PROGRAMME FOR WOMEN (STEP) was launched by Government of India in 1986 as one of the measures to ensure the well-being of women in the traditional and informal sector.

In Odisha, at present, the STEP Project is implemented by the Odisha State Cooperative Milk Producers Federation Limited (OMFED) in undivided district of Cuttack, Puri, Dhenkanal, Keonjhar, Sambalpur, Ganjam and Balasore.

Out of seven undivided districts of Odisha under WDP, two districts namely Cuttack and Dhenkanal were selected on the basis of performance, milk route and agro climatic conditions.

On the basis of milk procurement Cuttack district milk union leads among seven districts milk unions under WDP procuring 33,68,000 lts. milk against lowest procurement of Dhenkanal districts of 82,000 lts.

Similarly, in the area of average milk procurement and average milk procurement per DCS, undivided Cuttack district milk union occupies a leading position than other districts milk unions. Moreover, Dhenkanal districts milk union performance is the lowest among other districts milk union under study. With regard to organization of MPCS, Cuttack districts organized maximum up to 493 as against 161 numbers in Dhenkanal districts.

Keeping in view the location of these two districts in different agro climatic zone, there has been a wide gap with respect to performance of MPCS in general and WDCS in particular.

b) Selection of milk routes and WDCS

The two district milk unions named Cuttack milk union and Dhenkanal milk union were selected for study. Under each district milk union there are milk routes clustering a number of WDCS for easy operation of milk procurement and input supply activities.

Table 4.3 : WDCS and milk routes under Cuttack and Dhenkanal Milk Unions

Facts	Cuttack Milk Union	Dhenkanal Milk Union
Total no. of MPCS	493	161
General MPCS	278	79
WDCS	215	82
Female Members	21,492	6,930
Milk Routes	36	9

For deciding the sample size, from each district two milk routes, one having highest number of WDCS and other having lowest number of WDCS were selected. From each route fifty percent of the WDCS, were selected randomly for the study.

Table 4.4 : WDCS in highest and lowest milk routs under Cuttack and Dhenkanal Milk Union

Milk routs and WDCS	Cuttack Milk Union	Dhenkanal Milk Union	No. of WDCS selected
Name of the milk route having highest no. of WDCS	42-Mouza	Angul-Athamalik	-
No. of WDCS present in the highest route	18	20	19
Name of the milk route having lowest no. of WDCS	Chandipada	Kamakhynagar	-
No. of WDCS present in lowest route	2	4	3
Total	20	24	22

Table 4.5 : WDCS in highest and lowest milk route under Cuttack Milk Union

Milk Route	No. of WDCS	Sample size (50%)
42-Mouza	18	9
Chandipada	2	1
Total	20	10

Table 4.6 : WDCS in highest and lowest milk route under Dhenkanal Milk Union

Milk Route	No. of WDCS	Sample size (50%)
Angul-Athamalik	20	10
Kamakhynagar	4	2
Total	24	12

c. Selection of respondents

As per the objectives set for the investigation, from each WDCS two executive body members such as president and secretary were selected and three active members were selected randomly for collection of information . The details of selection of respondents are given below;.

Table 4.7 : Selection of respondents under Cuttack Districts Milk Union

Sl.No.	Name of WDCS	No, of President	No. of Secretary	Member	Total
1	Bantunia	1	1	3	5
2	Basantakudal	1	1	3	5
3	Chanduli	1	1	3	5
4	Dadhibabanpur	1	1	3	5
5	Kadampada	1	1	3	5
6	Maa Jagulei	1	1	3	5
7	Sisua Patna	1	1	3	5
8	Swapneswar	1	1	3	5
9	Varagola	1	1	3	5
10	Raghunathpur	1	1	3	5
	Total	10	10	30	50

Thus, a total of 50 respondents from ten WDCS were selected for the study under Cuttack districts milk union. Similarly the selection of respondents under Dhenkanal districts milk union is given below.

Table 4.8 : Selection of respondents under Dhenkanal District Milk Union

Sl.No.	Name of WDCS	No, of President	No. of Secretary	No. of Member	Total
1	Baradakota	1	1	3	5
2	Bhoipada	1	1	3	5
3	Deulabeda	1	1	3	5
4	Gautampur	1	1	3	5
5	Kanthapada	1	1	3	5
6	Karadasing	1	1	3	5
7	Kundanali	1	1	3	5
8	Malipadar	1	1	3	5
9	Odosingha	1	1	3	5
10	Sargiseni	1	1	3	5
11	Kadua	1	1	3	5
12	Mathakargola	1	1	3	5
	Total	12	12	36	60

Thus, from 12 WDCS under Dhenkanal districts milk union a total of 60 respondents were selected for study.

Table 4.9 : Category of respondents selected for study

SI. No.	Category of respondents	Cuttack Milk Union	Dhenkanal Milk Union	Total
1	No. of society selected for study	10	12	22
2	President selected	10	12	22
3	Secretary selected	10	12	22
4	Members selected	30	36	66

(Total number of respondents - 110)

6. Tools & Techniques used for Data Collection

a) Pilot study

Prior to preparation of the interview schedule, a pilot study was undertaken in Maa Jagulai WDCS functioning under Cuttack District Milk Union. The main objective was to collect first hand information about the procedure of functioning of the WDCS. A total no. of twenty respondents of which 2 executive body members were informally contacted and interviewed with the help of a preliminary interview schedule to gain the experience of preparing final interview schedules Such step, infact, helped immensely in preparing the final interview schedule.

b) Preparation of interview schedule

An interview schedule was developed as per the objectives set for the study. It contained both open and close-ended questions arranged in the sequence as per the objectives. It contains eight parts.

First part related to the socio-personal characteristics of women diary farmers on 15 variables such as age, education of respondent and her husband, family type, family size, religion, occupation, land holdings, herd size, milk production, consumption and sale, house type, animal shed, farm powers.

The second part of the schedule was related to collect data on procedure of formation and functioning of WDCS as per by-law in order to analyse the gap between the prescribed procedure and the procedure followed at WDCS level. This includes the variables like objectives of WDCS, funds, membership, general meeting, managing committee, qualification of members of managing committee, duties and functions of Chairpersons, duties and functions of the Secretaries and distribution of profits, election.

The socio-economic impact on the members of WDCS were studied in the third section of the schedule. The variables identified for study are; changes in pre-WDCS and post WDCS periods with respect to possession of buffaloes, improved buffaloes, local cows, cross breeds, changes in milk production , consumption and sale rate of milk, requirement of extra manpower were also studied under this section.

The data on change in materials possession such as bicycle, tractor, Pump set, T.V. and the area covered under fodder production were also collected in this part. Similarly, data pertaining to changes after being the members of WDCS, pattern of investment of extra-income from WDCS, extent of participation in extension activities were also collected for analysis and interpretation.

The 4th part of the schedule relates to the management activities of dairy animals in the WDCS under study. The variables identified for study are fodder cultivation, breeding practice, cost per breeding, feeding practice, frequency of vaccination, cost of treatment and handling of dairy management practices by the family members. In the fifth part of the schedule forty seven activities of the WDCS were identified and they were clubbed into nine key performance areas such as milk procurement, sale of milk and trend of payment, veterinary services, cattle feed and fodder, vaccination, breeding of dairy cows, credit facility of WDCS, training & extension and overall satisfaction.

Attempts were also made to study the entrepreneurial characteristics of WDCS members under the 6thpart of the schedule. The seventh part of the schedule dealt the problems faced by the members. Their suggestions for improvement of WDCS were recorded in the last part of the schedule.

c. Pre-testing of the interview schedule

After development of the schedule, pre-retesting of the schedule was done with 20 members in Maa Jagulai WDCS in order to assess reliability and validity of the schedule to record data. On the basis of the observation made during the pre-testing of the interview schedule, necessary modifications and refinement were made in the schedule.

d. Collection of data

One of the most important aspects of a quality research is the data collection. Personal contact was made by the investigator in interviewing the respondents selected for the investigation. Good rapport was established with desirable climate setting with the respondents thus helped the investigator to ask question and discuss on various issues pertaining to the study.

The purpose of the study was clearly appraised to the respondents before asking questions. Each aspect of the schedule was thoroughly discussed and appraised to the respondents. The informations received from the respondents were duly recoded then and there for analysis. The interviewer usually made nights halt in the villages to personally interview and record all information.

7. Operationalization and Measurement of Variables

7.1 Selection of Variables

Variable is a property that takes different values. Independent variable is the presumed cause of dependent variable. The independent variables included for the study were selected on the basis of extensive review of literature, discussion with experts and preliminary study conducted in the area of investigation. Only those variables which were found to have relevance with the objectives of the study were selected for detail investigation. The list of the variables are given below;

1. Age
2. Education
3. Family type
4. Family size
5. Caste/Tribe
6. Religion
7. Family Main occupation
8. Operational land holdings
9. Herd size
10. Milk production/consumption/sale
11. Type of house
12. Type of animal shed
13. Farm Powers
14. Procedure, formation and functioning of WDCS as per by-law.
15. Socio-economic impact on members of WDCS
16. Dairy managemental aspects such as fodder cultivation, feeding, breeding, vaccination and involvement of family members in dairy practices.
17. Key performance areas of WDCS
18. Entrepreneurial characteristics of WDCS.
19. Problems faced by the members of WDCS.
20. Suggestions for improvement of WDCS.

7.2 Measurement of Variables

Variables used in the study were quantified depending on their relative merit. In quantifying the measurement scoring method was adopted to measure the variables as detailed below;

i. Age

It is the chronological age of the respondents and spouse in completed years at the time of inquiry. The scoring made for analysis is as follows.

Age	Scoring
Up to 30 years	1
31 years to 50 years	2
Above 50 years	3

ii. Education

Education indicates that the level up to which the individual had studied. The number of years of formal education attended by the respondents were taken into account. Scoring of spouse education is also done in the same manner.

Education	Score assigned
Illiterate	0
Primary	1
Middle	2
High School	3
Intermediate	4
Graduate	5
Post-Graduate	6

iii. Family type

Family type was operationalised on the basis of whether the members were jointly staying with a common kitchen or not. The scoring is done as follows;

Family type	Score
Nuclear	1
Joint	2

iv. Caste

The term caste has been conceived as the endogamic, exclusive and functional group to which an individual belongs. The scores assigned to caste for the respondents as follows.

Caste	Score
SC	1
ST	2
OBC	3
General	4

v. Family occupation

Occupational status refers in the study are farming and other sources of income. The scoring pattern followed given below;

Occupation	Score
Wage earning	1
Family occupation	2
Farming	3
Dairy	4
Business	5
Service	6

vi. Land ownership

Land ownership is operationalised as to how much land is registered in name of the respondent or family members. The scoring is done as follows;

Land owner	Score
No land	0
Others	1
Self	2
Husband	3

vii. Fodder area

Fodder production is an important factor for profitable dairy farming. The fodder area is operationalised as to how much area the respondent utilized for fodder production. The scoring procedure adopted as follows;

Fodder area	Score
No fodder land	0
<0.5 hect.	1
>0.5 hect to 1 hect.	2
>1 hect.	3

viii. Herd size

Herd size in this study refers to the number of animals maintained by the respondents for milk production. Respondents usually keep deshi, crossbred and buffaloes for milk production. The score assigned for each as detail below;

Animal nos.	Score
1 - 3 nos	1
4 - 6 nos	2
7 and above	3

ix. Milk production

Milk production of the milch animals such as deshi, cross breed and buffalo was quantified on basis of the quantity of milk production per day. The frequency of range of production are calculated on the procedure of tally mark.

a. Milk production of deshi cow

Range (Lits.)	Score
1 - 3 lits.	1
4 - 6 lits.	2
7 lits. and more	3

b. Milk production of cross breed cow/buffalo

Range (Lits.)	Score
1 - 5 lits	1
6 - 10 lits.	2
11 Lits and more	3

x. Milk consumption

The respondents consume milk in their home. Such behaviour speaks about the health consciousness of the respondents. Milk consumption variable is operationalised on the basis quantity of milk the respondents consume in their family per day. The frequency of range of production are calculated on the procedure of tally mark.

Consumption Range (lt.)	Score
0.5 (Lit.)	1
>0.5 lit. to < Lit.	2
>1 lit.	3

xi. Milk sold to WDCS

The prime activity of the members of WDCS is the sell of milk to the society for making profit. The milk sold to the society refers to the quantity of milk a member sold to the society per day. The frequency of range of quantity of milk sold is calculated on the procedure of tally mark.

Range (lits.)	Score
<1 lit.	1
>1 lit. to 3 lits.	2
>3 lits. to 5 lits.	3
>5 lits.	4

xii. Shed structure

The structure of animal shed is an indication of the socio-economic soundness of the respondents. The animal shed structure is operationalised on the basis of the type of shed structure whose scoring has been done as follows;

Shed structure	Score
No house	1
Kuccha	2
Semi pucca	3
Pucca	4

xiii. House type of respondents

This refers to the type of house the members have for their residential purpose. This is operationalised on the basis of ingradients used in the construction of the house.

House type	Score
No house	1
Kuccha	2
Semi-pucca	3
Pucca	4

xiv. Situation of animal shed

Location of animal shed inside or outside of the residential structure is an indication of the knowledge of the respondent about the condition

Contd.

of animal shed. It is quantified on the basis of the frequency of responses of the respondent to different dimensions of the location. Such dimensions are scored as follows;

Dimensions	Score	Dimensions	Score
With in residence	0	Outside residence	1
Not adequate	0	Adequate	1
Not suitable	0	Suitable	1
Not satisfactory	0	Satisfactory	1

xv. Use of farm power

Use of farm power in farm operation indicates the state of farmers in using advanced technologies. More use of electric operated powers is an indicative of mechanization of farm operation and more use of bullock is an indication of traditional way of farming. Different means of operations are scored as follows;

Type pf farm power	Score
Bullock	1
Buffalo	2
Electric operated power	3
Power tiller	4
Tractor	5
Not cultivated	0

xvi. Objectives of WDCS

In order to identify the gap in the process and procedures followed in formation of WDCS prescribed by the concerned authorities, the by-law was referred. The objectives starting from 01 to 05 were asked to the respondents to document whether they are followed or not followed. The scoring is done as follows;

Objectives	Score
Completely followed	2
Partially followed	1
Not followed	0

However, for adopting the strategies for raising funds of society, criteria for inducting members, adhering to the qualifications of members for managing committee and election procedure, the score '1' was given for followed and the score '0' was given for not followed.

With respect to the duties and functions of the chairperson and the secretary, the opinions of the respondents are secured with the help of a three point scale scored as below;

Statements	**Score**
Fully satisfactory	2
Satisfactory	1
Not at all	0

Similarly, the distribution of profit of the society was operationalised on the basis of extent of management of distribution among the members. The opinions of the respondents were collected with the help of a three point scale and quantified with the help of tally mark. The scoring was done as follow;

Statements	**Score**
Well managed	2
Managed to some extent	1
Mismanaged	0

xvii. Change in possession

Change in possession refers to the quantitative difference in different dimensions such as dairy animals, material possession, livelihood, investment and capacity building during pre-WDCS and post-WDCS period. This was operationalised on the basis of counting frequency of the change occurred on possession of dairy animals after being the member of WDCS. For counting frequency scoring was done as below;

Situation	**Score**
Change	1
No change	0

Similarly change in livelihood aspects and investment pattern of the member of WDCS were operationalised on the basis of assigning scores ;

Response	**Score**
Yes	1
No	0

The quantification of the responses were made as per the procedure of tally mark. In the case of change in capacity building, each participation was scored '1' and non participation was scored '0' and hence, quantification is made on the basis of frequency of responses.

xviii) Growing of green fodder

Attempt was made to know whether the WDCS members go for fodder cultivation or not . This variable was operationalised on the basis of dichotomus questions. The responses were quantified as to how many responses were under 'yes' or 'no' category. However, for quantification scores were allotted '1' for Yes '0' for No.

xix) Practices of breeding, feeding and vaccination of animals

It means the type of breeding, feeding and vaccination practices adopted in the animals by WDCS members. This was operationalised on the basis of assigning scores to the extent of adoption of breeding, feeding and vaccination practices as follows :

Frequency	Score
Always	2
Sometimes	1
Never	0

xx) Sources of cattle feed

It means the sources from where the WDCS members purchase the cattle feed such as concentrate, green fodder and roughagc. This was operationalized on the basis of choices of source they opted to purchase by assigning the following scores ;

Sources	Score
WDCS	3
Local traders	2
Others	1

xxi) Participation in dairy management practices

The respondent and her family members involve in carrying out different dairy management practices in there home. An attempt was made to document the individual involvement in respective dairy activities. Altogether sixteen number of dairy management practices were identified and they are grouped under 5 important dimensions such as ;

i) Involvement in feed, fodder, feeding and watering activities

ii) Involvement in breeding practices

iii) Involvement in making access to treatment

iv) Involvement in cleaning and sanitation of cow-shed

v) Involvement in Milking and Milk Marketing.

Six alternatives among the participants such as the respondent herself, her husband, servant, herself and husband, husband and servant, were chalked out in order to know their respective involvement in each activity. Score '1' was given for each participation in a specific activity and score '0' for non-participation. Quantification was made on the basis of number of frequency of participation or non-participation.

xxii) Key performance areas of WDCS

Each WDCS works under the guideline of it's by-law approved by State Government. Moreover, the WDCS has some specific function to do. The duties and functions of WDCS were (detailed from the by-law and a total 47 variables were identified for study. All such 47 variables were clustered into 9 different key performance areas on the basis of their similar nature of activities.

1. **KPA_1 : Regularity in Milk Procurement of WDCS**
    1. Procurement of milk is regular
    2. Milk procurement is in increasing trend.
    3. Milk procurement is non stop
2. **KPA_2 : Sale of Milk and Trend of Payment**
    1. Regularity in pourer
    2. Regularity in payment
    3. Selling milk to local consumers
    4. Milk sold outside
    5. Sale of milk products
    6. Remunerativeness of milk products
3. **KPA_3 : Availability of Vety. Services**
    1. Meeting of emergency services
    2. Availability of Vety. Doctors
    3. Fixtation of fee of Vety. Doctors by WDCS
    4. WDCS equipped with first aid equipments

5. Bringing cows for WDCS for first aid
6. First aid on door-step service is provided

4. **KPA_4: Feed and Fodder Activities of WDCS**
    1. Purchase of cattle feed from union
    2. Purchase of cattle feed from outside
    3. Growing of green fodder
    4. Availability of planting materials for green fodder
    5. Growing of fodder throughout the year
    6. Knowledge about fodder cultivation
    7. Satisfaction about yield of green fodder.
5. **KPA_5: Vaccination Activities**
    1. Regularity in vaccination
    2. Vaccinating all animals in time
    3. Purchase of vaccines from union
    4. Purchase of vaccines from outside
    5. Vaccination at free of cost
6. **KPA_6 : Breeding Facility of Dairy Cow in WDCS**
    1. AI is timely and regular
    2. Making use of breeding bulls
    3. Availability of required number of breeding bulls
    4. Maintenance of breeding bulls by the society.
7. **KPA_7: Credit Facility at WDCS**
    1. Easy access to bank credit facility
    2. Higher rate of interest of commercial bank
    3. Complicated procedure to avail credit.
8. **KPA_8 : Training and Extension Programmes Conducted by WDCS**
    1. Training programme is regular
    2. Training is need based
    3. Training facility is at free of cost
    4. Regular dairy extension programmes

5. Satisfaction of WDCS members on extension service
6. Need based extension programme
7. Participation of members in extension programme

9. KPA_9 : Overall Satisfaction by WDCS Members

1. Satisfaction on milk sale
2. Satisfaction on quality and price of milk
3. Satisfaction on veterinary service
4. Satisfaction on quality of vaccine
5. Satisfaction on training programme
6. Satisfaction on AI services

In order to quantify the response, each response is scored '1' for agreement and '0' for non-agreement to the statement. Thereafter, all the responses were counted and calculated with the help of tally marks and put under statistical methods for analysis.

xxiii) Entrepreneurial characteristics of WDCS members

In an attempt to study the entrepreneurial characteristics of WDCS members ten entrepreneurial dimensions were identified to measure the gap in entrepreneurial behaviour of members of WDCS. These dimensions are, hard work, desire of achievement, optimistic, foresightedness, independent, future planning and organizer, innovativeness, information seeking behaviour and quality consciousness. Each response was measured on a three point scale with assigning scores as follows :

Statement	Score
Agree	2
Disagree	1
Do not know	0

xxiv) Problems faced by the members of WDCS

Despite of provision of facilities to support the members of WDCS, the members faced some problems. Eleven such problems were identified and operationalised on the basis of weightage given to the problems by each member of the WDCS. Accordingly, problem ranking was prepared starting from most prevailing problem to least prevailing problem.

xxv) Suggestions for improvement of WDCS

Attempts was made to operationalize the suggestions of WDCS members by identifying seven important areas on which the members suggested for further improvement. Under each dimension of problem some important suggestions were identified and sought their response to each suggestion with the help of dychotomus questions to answer in 'Yes' or 'No' form. All such responses pertaining to each dimension was calculated with the help of tally marks and analysed statistically.

7.3 Data Processing and Statistical Methods Used

The following statistical methods were used in this study for analysis of responses based on the nature of the data and type of information required.

1) Percentage

Percentage was used in descriptive analysis for making comparisons. To calculate the percentage, the frequencies on the particular cell was multiplied by 100 and divided by the total number of respondents in that particular category.

ii) Mean score

It is also another simple comparison, which is calculated by using the formula.

$$M.S. = \frac{\sum fx}{N}$$

Where $\sum fx$ = Total score obtained in individual category.

N = Total number of respondents

iii) Rank order analysis

On the basis of average mean score, the rank order was made. The item scoring highest mean score was given first rank and the next highest with 2nd rank and so on.

i) Correlation

It was employed to find out the association of the dependent variables with the independent variables. It was also used to observe relationship of the independent as well as dependent variables themselves. Parson's coefficient of correlation was used to find out zero order correlation between dependent and independent variables to see the degree of closeness of linear relationship. The formula used for calculating co-efficient of correlation is as follows.

$$r = \frac{N\sum xy - (\sum x)(\sum y)}{\sqrt{N\sum x^2 - (\sum x)^2 \times N\sum y^2 - (\sum y)^2}}$$

Where N = Number of pairs correlated

x& y = The variables being correlated

$\sum$ = Summation of overall cell entries of variables

r = Correlation of co-efficient between x & y

ii. Test of significance

This test was used to know whether the observed sample correlation is statistically significant to indicate correlation in the population. The test is implied by using the formula.

$$T = \frac{r}{\sqrt{1 - r^2}} \times \sqrt{N - 2}$$

T = Calculated value of statistics student 't' from the sample observation which has N-2 degrees of freedom.

R = Calculated correlation value

N = Number of pairs correlated

If calculated 't' value exceeds the table value either at 5 % or 1 % level of significance, then correlated value is accordingly significant at 0.05 or 0.01 level of probability.

iii. Critical ratio test

The method is used to test the significance of the difference between percentage. The test is implied by using the formula as given below.

$$P = \frac{N_1P_1 + N_2P_2}{N_1 + P_2}$$

$$Q = 1\text{-}P \cong 100 - P(\text{Value})$$

$$\sigma D^2\% = \sigma P_1P_2 = \sqrt{\sigma^2 P_1 + \sigma^2 P_2} = \sqrt{PQ + \frac{1}{N_1} + \frac{1}{N_2}}$$

$$C\text{-}R = \frac{(P_1 - P_2) - 0}{\sigma P_1 - P_2}$$

Where

C.R.	=	Critical ratio value
N_1 & N_2	=	Size of both sample categories
P_1 & P_2	=	Percentage occurrence of a given behaviour
P	=	Percentage occurrence of the pooled behaviour
Q	=	Differential percentage of the pooled behaviour
AD^2 %	=	S.E. of the difference between P_1 and P_2

❑❑❑

Chapter - 5

Functioning of Women Dairy Cooperative Societies

The functioning of the Women Dairy Cooperative Societies depends upon the variations in managerial capacity leading to profit and loss of different WDCS. To make an in-depth analysis, the study was undertaken in undivided Cuttack and Dhenkanal district of Odisha covering 22 randomized WDCS with a sample of 110 women respondents.

The analysis of the findings after statistical treatment have been presented objective-wise as follows;

i. Socio-economic profile of the members of WDCS involved in dairy production activities.

ii. Gap between procedures prescribed and followed in formation and functioning of WDCS.

iii. Identification and analysis of key performance areas in the WDCS.

iv. Determination of management practices and milk production profiles of WDCS members.

v. Extent of participation of women members in dairy activities.

vi. Entrepreneurial characteristics of members of Women Dairy Cooperative Society.

vii. Impact of WDCS on socio-economic status of the members.

viii. Correlating variables and their interrelationship in predicting smooth functioning of WDCS.

ix. Constraints faced by the WDCS members.

x. Suggestions of members for improvement in functioning of WDCS.

I. Socio-Economic Profile of Members of WDCS

Co-operative exists on human interaction. The interaction among the individuals laid the foundation of group dynamics. The present concept of development revolves around the social capital, economic capital and information capital. The social capital essentially includes profile of informants in which they reside and work, since the co-operative society is based on the concept of group dynamics, it was necessary to investigate into socio-economic profile of members of WDCS.

In Odisha, at present, there are 918 numbers of WDCS with membership to the tune of 68,851. The members of WDCS come from different strata of society and hence exhibit difference in their achievement. Social studies have proved beyond doubt that performance of any cooperative effort relate to socio economic profile of the participants. Keeping these hypotheses in view the present study examines as much as 15 variables under socio-economic profile as has described below;

1. Age of the members of WDCS

Chronological age of an individual is related with the activity aspiration, interest and future plan of action. It is hypothesized that individual of middle age groups are more sensitized to new activities and venture for new avenues for livelihood.

Data reveal that majority of the respondents (62.72%) belong to age group of 31-40 years and 40-50 years of age. However, there were only 9.1 % of respondents who crossed the 50 years of age and up to 30 years their percentage was 28.18%. The result as a whole reveals that most of the members of the WDCS are middle age group. Therefore, their response can be taken as valid and reliable on the basis of their experience.

2. Education of the members of WDCS

Education is an important component for development and that too for sustainable livelihood. The education has been attached with much importance for climbing up socio economic ladder. The % of literacy of women in Odisha is reported to be 50.97 % (2001 census).

Of the sample 17.27% were Illiterate, while 1.84 % had the opportunities to graduate level. The other distribution reveals that 33.63 % were educated up to primary level, 17.27% middle school, and 23.63 % up to high school level. Comparatively members of WDCS of Cuttack district are little more educated than Dhenkanal district.

3. Spouse education

The study also considered spouse education as an important determinant in assigning socio economic status. The educational qualifications of their counter part have been taken into consideration.

The data reveal that of the sample, 7.3% of their spouses were illiterate where as one had an opportunity to avail post graduation qualification. Of them 29.09 % were educated up to primary level, 27.27% up to high school, 20.90% up to middle school and 10% up to intermediate level. The spouse education appears to be better than education of the sample members.

4. Family type

The composition of family bears direct or indirect influence on the activities of the individuals of the family. Results reveal that 57.27 % belong to nuclear family system against joint family system of 42.73%. The findings coincide with the observations of many authors who worked on demographic aspects of ward type.

5. Family occupation

In Odisha family occupation still continues to be a determinant of livelihood system. The family occupation is also related to socio-economic status of the people. With change of time there has been diversification, but influence on family occupation continues to exist.

The data reveal that of the sample, majority of the respondents (71.81 %) depend upon farming, followed by dairying (29.09%), and business and service occupy an equal level that is 18.18%. However, 10% of the sample used to depend upon wage earning and 5.45% in caste occupation.

6. Trend of caste occupation

In Odisha and other parts of the country, caste is still considered as an important factor in livelihood system. The State as well as National Govt. has classified caste into SC, ST, OBC & General.

A glance on the data reveals that 58.19 % belong to general category and 29.09% to OBC. The percentages of sample belong to SC & ST was 8.18 and 4.54 respectively. The caste distribution appears to be the same in both the districts.

7. Land ownership

Land possession of a family in Odisha is considered from social, economic, cultural and political point of view. Normally the land is recorded in the name of the family head mostly male members which is witnessing change in recording of ownership of the land.

The analysis of data reveal that the land is recorded in the name of the husband and other members of the family to an extent of 35.45% and 50% respectively. The samples under study were found to be 6.36% as having ownership on land. However, 8.19% of the sample did not posses the land.

8. House structure

Residential house, it's structure, size and composition have included as the indicators for assessing socio-economic status. With change in time and style of living, there has been changes in house structure in rural areas of Odisha.

The study attempted to find out residence house structure of the members. Of the sample equal percentage (28.18%) had pucca and semi-pucca houses where as majority of them (43.64%) had residential unit of kaccha structure. In other words, it reveals that still a high proportion of house hold reside in kaccha house.

9. Land under fodder cultivation

The present study is designed to investigate into WDCS dealing with milk and milk products. It was felt necessary to investigate into the land under fodder cultivation of the members. It is so because green fodder is considered to be an indicator for success of dairy units.

It is interesting to note that, of the sample as high as 88.20% had no land under fodder cultivation, where as 9.09% had less or equal to 0.5 hectare, 1.81% more than 1 hectare, 0.90 % had 0.5 to 1 hectare of land under fodder cultivation. The results present a discouraging situation with regard to area under fodder cultivation. It needs intervention of concern authority.

10. Use of farm power

The agricultural operation of the state in general is tending towards mechanization. The use of animal power is generally being replaced by machines owing to various reasons. While finding out the situation of farm power, the following results were obtained.

The data reveal that, of the sample 65.45 % depend upon bullock for farm operations, 8.18 on buffalo. Tractor has considerably gained momentum in rural sector upto 30.90% and power tiller only up to 1.81%. Still in rural areas the farm power scenario reveals animal power for cultivation.

Comparison of Socio-economic Status of the Members of WDCS in Cuttack and Dhenkanal Districts

It was necessary to find out difference between Cuttack and Dhenkanal district as the members exhibited differential behaviour to socio-economic indicators. Applying suitable scoring methods, the analysis was made to find out difference of two districts as has been shown table below.

Table 5.1 : Comparative socio-economic status analysis of members of WDCS belonging to Cuttack and Dhenkanal district

Sl.No.	Indicators	Cuttack	Dhenkanal	Difference (%)
1	Age	1.98	1.66	16.16 *
2	Education	1.98	1.53	22.72*
3	Sp. Education	2.62	1.86	29.00*
4	Family type	1.46	1.33	8.90 NS
5	Family Occupation	6.60	4.63	29.84*
6	Caste composition	3.56	3.21	9.83 NS
7	Land Ownership	1.66	1.71	2.92NS
8	Fodder Area	0.08	0.23	65.21*
9	House Structure	2.34	1.43	38.88*
10	Farm power	3.62	1.45	59.94*
	Average	2.59	1.90	28.57

N.S : Not significant * Significant difference

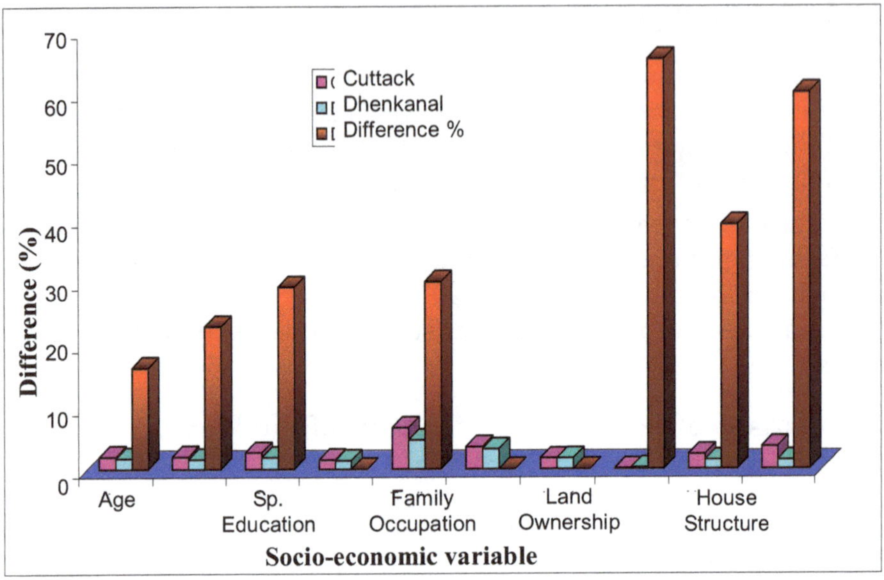

Fig. 5.1 : Comparative socio-economic profile of the members of WDCS in Cuttack and Dhenkanal districts

The score analysis of data on 10 indicators presented in the table above revealed that two districts do not differ significantly in matter of land ownership, caste composition and family type where as they differ on other variables significantly. The highest difference is marked in case of area under fodder cultivation followed by possession of farm power, house structure, family occupation, education of spouse and self. Taking total difference into account the difference between Cuttack and Dhenkanal districts is found up to 28.57%. The district Dhenkanal occupies lower position in all respects except land ownership, area under fodder cultivation. Since the district Dhenkanal comes under mid central table land zone comprising of more of rocky and high land, the extent of cultivation of fodder is found to be higher. Moreover, considering the factors like age, education, education of spouse, family occupation, house type and farm power Cuttack occupies first position because of progressiveness, more facilities, urbanization and influence of higher percentage of literacy of the district.

II. Gap between Procedures Prescribed and Followed in Formation and Functioning of WDCS

As per the Act of the cooperative society Women Dairy Cooperative Societies in Odisha were established in accordance with prescribed rules and regulations formulation by Central Cooperative authority and approved by Government of Odisha.

The central cooperative society dedicated to the cause of rural milk producers was taken into considerations. The present investigation examined the procedure, formation, functioning of WDCS with respect to ;

(1) Objectives

(2) Funds

(3) Membership

(4) General meeting

(5) Management committee

(6) Qualification of members

(7) Duties and responsibilities of chairperson

(8) Duties and responsibilities of secretary

(9) Distribution of profit

(10) Election procedures

The above dimensions under study were qualitatively and quantitatively examined following appropriate scoring and scaling technique.

1. Objectives of women dairy cooperative society

The establishment of the WDCS is based on sound policy of the State Government to encourage the milk production of the state. The Odisha State Cooperative Milk Producers' Federation is the authorized agency to regulate WDCS with respect to duties and responsibilities to be discharged.

The present study is investigated into specified objectives of WDCS to examine whether the activities undertaken by sample are in confirmative with the prescription. The collected information under score analysis revealed that the overall gap in understanding the objectives of WDCS is 32 per cent. Comparing the two districts, gap is found more in case of Dhenkanal district followed by Cuttack district. Out of the five dimensions of the objectives highest gap is observed in case of fodder cultivation (62.5%) followed by improvement of breeds and health of animals (35%), sale of nutritional balance cattle feed (26%), production of quality milk (18.5%) and making milk marketing profitable (18%). The results as a whole are indicative of the fact that emphasis has not been given to understand the objectives of WDCS fully and specifically with respect to encouragement of fodder cultivation. The general observation for two districts is in question. The milk producers mostly depend upon paddy straw and concentrate rather than focusing on fodder cultivation.

2. Funds

For operation of society there is need of funds. The funds have to be raised from the members or to be supplemented by the State Government. In analyzing the fund raising system of the WDCS under study the data reveals that fund mobilization of the WDCS is found 100% on issue of share followed by membership fee 95.45%. It is interesting to note that 30.90 % expressed of receiving donations for operation of WDCS followed by receiving deposits up to 29.09%. However, incurring of loan is found to be 17.27% of the respondents. The findings as a whole revealed that issue of share and membership fee are the major source of finance to start a WDCS in the area of study.

3. Membership

Enrollment of the members is the major objective of any cooperative society. It is hypothesized that more the members, greater is the progressiveness of the society. But to be a members of the WDCS there are certain guidelines as follow. These have been framed for smooth operation of the society .

1. Residence within operational area.
2. Making application in prescribed form.
3. Having buffalo/cows for milk production.
4. Supply of milk to WDCS for three consecutive months.
5. Payment of membership fee.
6. Written agreement to supply milk to society only.
7. Not declared bankrupt.
8. Not convicted for any crime.

While examining the membership drive of the sample societies the following results are obtained, through score analysis method.

The analysis of the data explains that all the members of the respective society do live within the operational area of it. The members of the society agreed to make supply of milk to WDCS for three consecutive months followed by making applications to enroll themselves as the members of society in prescribed form and make agreement to supply milk to the society. These are the minimum requirements to become members. The deviations found (0.64) in case of payment of membership fee, followed by having buffalo and cow for milk production (0.73) and not declared as bankrupt (0.92). However, the findings as a whole reveal that the members of WDCS satisfy most of the requirements to be the members of the society. Taking membership as a whole the deviation is observed to be maximum of 11% which is permissible incase or rural areas of Odisha. It shows that while organizing the WDCS care has been taken to select members having eligibility and other considerations.

4. General meeting

It is mandatory for the society for organizing annual general meeting within a period of three months, elect chairperson, management committee, sanction of budget ,making necessary arrangements for collection of the milk and recording of proceedings. The normal rules of the WDCS are to be acted upon for smooth operation of the society. While investigating into the matter the following results were obtained.

The result reveals that out of six functions of the general meeting 99.90% arrangements are made for collection of milk followed by 86.36% to elect the chairperson and 81.81% cases the members of the management committee declared. The annual meeting is held within a period of three months after the end of each cooperative year is expressed by 75.45% of the sample. Equally 82.72% mentioned that the records are maintained for each proceeding. While comparatively much less response (53.63%) was received in case of sanctioning of budget for the next year. A critical look at the analysed data reveals that 78.47 % of the sample opined that the regular meetings of the WDCS are held whereas 9.43 % opposed the view, living 12.10% who expressed their ignorance about the meetings. In short, about 21.49% are not in favorable opinion against the majority of 78.57 %. In other words, there is a good proportion of members who are not satisfied with the meetings of the WDCS to discharge the responsibilities as expected.

5. Management committee

Each and every cooperative society constitutes managing committee as a part of requirement for its smooth operation. As observed, in many cases the management committee does not function well owing to conflict and irregularities. The present investigation investigated the regularity of the management committee of sample of WDCS which yielded the following results.

The analysis of the data reveals that out of the total respondents of Cuttack district 12 % viewed management committee as most regular whereas 70% viewed as regular and 18% who opined as not regular. In case of Dhenkanal district not regular is viewed by 25% while regular and most regular is expressed by 60 % and 15% respectively. In other words, majority are in view that the management committee is regular. It is also ascertained that all the WDCS under study have management committee. But Management itself is running as per by-law or not remains to be examined.

6. Eligibility of management committee members

For enrollment of the members into WDCS, the Government of Odisha has stipulated guidelines to help in selecting the right members to the society. But in many cases the procedure is not followed owing to various reasons

While examining the qualification of management committee members of WDCS, it is revealed that the criteria like, not disqualified by legal proceedings and not found liable to any legal offence of the society are being followed up to the level of 95 % in both the cases followed by criteria does not bear direct or indirect interest (93.33 %), checking eligibility of the members and check whether defaulter on dues to society were 91.66 % in both the cases. However, the overall finding says that the members of WDCS followed the eligibility and qualification of management committee members to a satisfactory level.

The hypothesis of the study is that better the guidelines are followed higher is the rate of out put. The examination of the hypothesis study under score analysis revealed the following results.

Table 5.2 : Eligibility of the management committee members of WDCS (score analysis)

Qualification	Cuttack	Dhenkanal	Pooled
Checking eligibility of the members	1.00	0.91	0.95
Check whether defaulter on dues to society	0.96	0.91	0.93
Does not bear direct or indirect interest	1.00	0.93	0.96
Not disqualified by legal proceedings	1.00	0.95	0.97
Not found liable to any legal offence of the society	0.98	0.95	0.96
Supply milk regularly to the society	0.98	0.83	0.90
Any of the family members does not carry business to the society	1.00	0.76	0.88
Average	0.8	0.89	0.93

A glance at the table reveals that in many cases of Cuttack district, the eligibility criteria of members has been duly examined, but in case of Dhenkanal district there is deviation in supplying milk regularity to the society and family members does not carry any business to the society. The overall picture depicts that the deviation is up to 0.93 that is up to 7%. It is natural that while forming society the qualification and eligibility of members are examined as per norms but subsequently the members do not perform their responsibility as has been evidenced in case of regular supply of milk to the society and carrying out business with the society.

7. Duties and function of chairperson

Each and every WDCS has a chairperson who is elected by the *bona fide* members of the society. Their duties and responsibilities are directly proportional to the achievement of the society. It has been reported by a number of studies that the role of the chairperson in cooperative society act is the guiding force for achievement. The sample WDCS under investigation with respect to duties and responsibilities reveal the following results.

A look at the findings reveal that, of the total duties of chairperson, the maximum gap is found in case of deposit in and drawing of cash from the bank followed by taking action to rectify objections raised by the audit, furnishing information to Additional Registrar, regularity in supervising work of the society and execution of resolution passed by the Managing Committee in order. Comparison of the two district shows that maximum gap is found in Dhenkanal district (50%) as against 23% gap in Cuttack district. The overall gap of all dimensions together explaining the duties and functions of the chairperson is found 36.50%. In other words, more than $1/3^{rd}$ of the chairpersons are not performing their duties and functions as required by the regulation of the WDCS. Efforts are needed to educate the chairpersons about their duties and functions on functioning of the WDCS.

8. Duties and responsibility of secretary

In a cooperative society, the Secretary is the key person to regulate the function of the society in accordance with by-law. discharge the duties. Secretary should be aware of their duties, responsibility and limitations. The hypothesis of study is that more the Secretary is duty bound and functional, greater is the achievement. While examining the duties and function of Secretaries of sample WDCS, the data reveal that there is considerable gap in collection of money and deposit in bank, maintaining of cash book and normal records, placing audit report before managing committee in time withdrawal and recovery of money as per instructions of chairperson, receiving of communication and disposal of letters, preparation of receipts, vouchers and reports and maintaining of routine administration of society in order. However, minimum gap is observed in case of arranging of general meeting, management committee meeting and others. The overall gap is found to be 33.50%. The gap is found more in Dhenkanal District (42%) than Cuttack district (25%).

The data as such explained that majority of Secretaries are aware of almost two-third of their duties which do not permit the society to achieve their target. There is need for intervention to regulate collection and deposit of money in bank and maintenance of cash book and other records to activate the functions of WDCS. Such finding has been reported by many of the earlier investigators on Co-operative Societies.

9. Distribution of profit

The main attraction for joining in cooperative society to share profit and risk. Observation reveals that most of the conflicts start with distribution of profits among the members. The situation with the present sample of WDCS in equi-distribution of profit was found.

Analysis of the data revealed that, of the sample as much as 41.8 % expressed WDCS to be well managed followed by 32.72 % viewed management to some extent while one-third of the sample (25.47%) expressed profit is mismanaged. Comparing two districts, Cuttack is found in better position over Dhenkanal in distribution of profits among the members. The crucial intervention is suggested to be taken care of equi-distribution of profit as it serves as root cause of conflict.

10. General election of society

Election of Society is a regular feature as stipulated by State Govt. under Cooperative Act. The Bye-laws enforce to conduct election following notification, declaration of vacancy, nomination, voting, counting and declaration of results. Theses norms are to be followed for each and every society. An examination of election procedure of the sample WDCS revealed that all the election procedures were followed maximum up to 54 to 57%. In other wards there is great deviation in publishing notice for meeting, voting, declaration of results, filling of nomination, and notification of electoral roll. The above six dimension are integral part of electoral procedure. The WDCS fail to adopt fully the election procedure resulting problems of dissatisfaction, mismanagement and low output. The intervention on the part of authority is essential to remove these deficiencies.

11. Summarization of objective II

Gap in procedures followed in formation of WDCS was investigated on ten dimensions like objectives, sources of funding, membership pattern, conducting of general meeting, qualification of members, duties and responsibilities of chairpersons, duties and functions of Secretaries, distribution of profit and holding of election. The gap found in all the case are summarized below:

Table 5.3 : Ranking of gaps between procedures prescribed and followed in formation of WDCS

Sl.No	Dimensions	Gap (%)	Rank
1	Holding general election	43.94	1st
2	Duties & responsibilities of Chairperson	36.50	2nd
3	Duties & responsibilities of Secretary	33.50	3rd
4	Understanding the objectives of WDCS	32.00	4th
5	Sources of funding	30.90	5th
6	Distribution of profit	25.47	6th
7	Management committee meeting	21.82	7th
8	Conducting of General meeting	17.28	8th
9	Qualification for membership	07.00	9th
10	Membership drive pattern	05.50	10th

The gap analysis revealed that is case of all entire ten dimensions, gap varies from 5.50 % to as much as 43.94%. The highest gap is observed in case of holding general election, duties and responsibilities of chairperson, secretary, understanding objective of WDCS, sources of funding and distribution of profit. However, a negligible gap exists in case of membership drive, qualification for membership and holding of general meeting. The overall picture is not very much satisfactory.

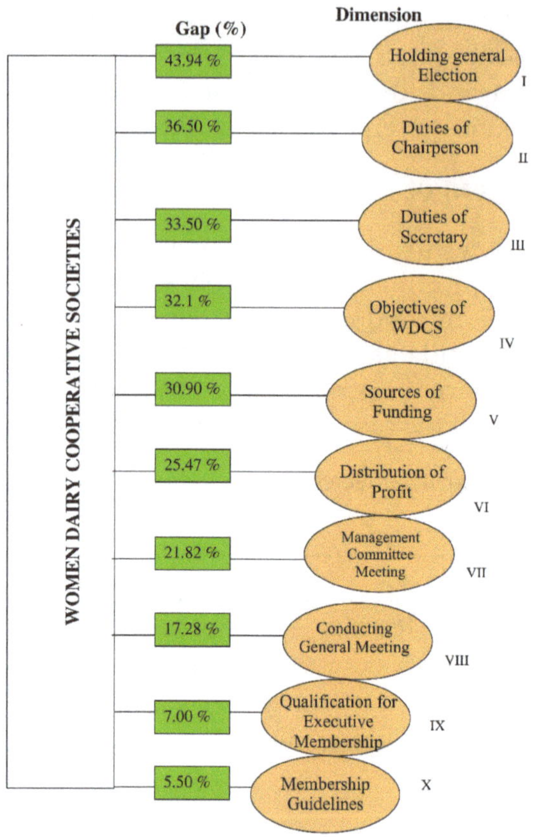

Fig. 5.2 : Ranking of gap between procedures prescribed and followed in formation of WDCS

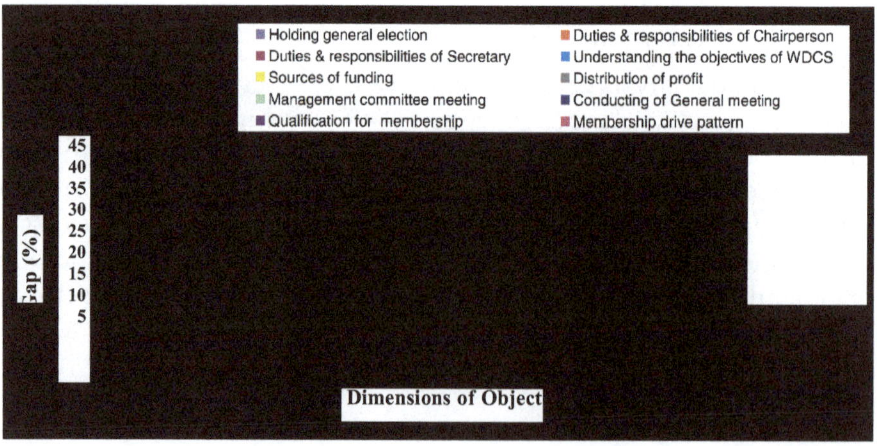

Fig. 5.3 : Ranking of gap between procedures prescribed and followed in formation of WDCS

III. Key Performance Areas of WDCS

WDCS are the key functionaries at village level to boost up milk production. The Government of Odisha followed Anand Pattern to establish WDCS to increase milk production and productivity thereby increasing household income level and bringing prosperity in the rural sector. At the beginning the participants and general public were very much impressed upon the success of WDCS. WDCS operates within certain norms and activities. The activities are called 'Key Performance Areas' (KPA). The study classified the key performance areas into nine dimensions such as milk procurement, sale of milk and payment, veterinary services, cattle feed and fodder, treatment and vaccination, breeding facilities, training and extension programmes and overall satisfaction of members of WDCS. These nine key areas have been studied and analyzed as described below.

1. Milk procurement

WDCS depend upon regular milk procurement and payment to the members of the society. Deviation in the milk procurement programme creates problem in the smooth operation of the society. The opinion of the sample respondents regarding regularity of milk procurement by the WDCS members were obtained as shown below.

The percentage analysis reveals that 94.54 % of respondents are in the opinion that milk procurement is regular in their WDCS, 80% viewed procurement is in increasing trend, while 71.81% viewed milk procurement is a non stop process in their WDCS. Comparing both the districts, regularity in procurement of milk, its increasing trend and procurement without break is almost higher in case of Cuttack district except the last determinants. In other words the majority of the sample expressed that procurement of milk of the WDCS members is very regular and that too increasing trend. Therefore, this aspect is to be taken as positive in the matter of progress of the WDCS.

2. Sale of milk and trend of payment

The key performance areas of the WDCS essentially include milk procurement, sale of milk and receive of payment. These are the basic activities of the WDCS to keep them growing. With sample respondents, about disposal of milk and payment was studied as furnished detailed below. A look at the findings reveal that regular payment is reported to be 95.45% of the sample followed by regularity in pourer (92.72%). Sale of milk product is reported to be lowest (20%) next to milk sold outside by the samples (32.72%). About 47.27% viewed that milk products are quite remunerative whereas 48.18% reported that they sell milk to local consumers at satisfactory price. Comparing the two districts under study, similar results are found in both cases. In short there is evidence that the members of the WDCS dispose their milk to the local consumers although they report that they contribute regularly to society. Therefore, an intervention is necessary to procure entire quantity of milk produced by the members of the society.

3. Veterinary services

Dairy cooperative society is intimately associated with the veterinary services. In Odisha there are 540 numbers veterinary hospitals with qualified veterinary doctors of 613. They provide veterinary services to the milk cooperative society members as and when required. The opinion of samples regarding veterinary services has been examined and presented below.

The results contained in analysis reveal that members avail emergency services as and when required to an extent of 92.72% followed by first aid service at the doorstep 67.27%. The members also expressed that they bring cow to WDCS for first aid service (49.09%) and availability of veterinary doctors is 43.63%. It is reported by 40.90% respondents that WDCS are equipped with first aid equipments. However, 31.81 % expressed that fee for veterinary doctors are fixed by the WDCS. Both districts did not defer much so far as availing of veterinary services is concerned. The results reveal that the members of WDCS were not fully covered in availing the services of veterinary doctors, equipping WDCS with first aid equipments and receiving first aid service at door step. It is also required that fee for veterinary doctors are to be standardized irrespective of the location of WDCS.

4. Cattle feed and fodder

The feed and fodder for animals are the prevailing problems for the state government. To achieve success in dairy sector there is a need to focus on availability of feed and fodder at the affordable rate to the members since feed and fodder are of prime importance. The study investigated in situation as revealed below.

The content of the findings reveal different dimensions of feed and fodder aspect. As much as 85.45 % reported of purchasing cattle feed from union where as 67.27 % from out side in addition to the union.

Grazing of green fodder is reported by 75.45%. The availability of planting material, knowledge of fodder cultivation and growing of fodder throughout year were reported by 63.63, 33.63, and 18.18% of respondents respectively. However, 41.81% reported that they do not harvest expected yield of fodder grass in the land. Similar trend about feed and fodder is observed in both the districts. The findings are in view that there is need to popularize fodder cultivation, imparting knowledge about growing of fodders and efforts are to be made to increase the yield per unit to sustain the milk production of the WDCS.

5. Vaccination

The general diseases in cattle of Odisha are FMD, BQ, HS etc. which reduce the milk yield to a significant level. Efforts are made constantly to keep animals free from these diseases. To avoid occurrence of diseases vaccination is recommended against different diseases according to the status of animals.

The reaction of the sample reveals that regularity in vaccination is followed up to 80% and most of them (80.90%) purchase vaccine from union through their respective societies. The response regarding vaccinating of all animals is reported by 59.09% respondents where as availability of vaccine at free of cost is expressed by 64.54%. Both districts exhibited same trend in case of vaccination, it's availability and use.

The study indicated that purchase and cost of vaccine needs to be standardized and uniformed so that all the members of WDCS avail equal advantage and benefits. The concerned authority needs to look into this matter in order to dispel disparity among the members of the society.

6. Breeding of dairy cows

The animal breeding for sustaining the level of production of milk is recommended by the state department as well as the research centers. The regular phenomenon of breeding is providing of AI services, making use of breeding bulls with sufficient numbers and maintaining them at society level for equal benefits of the members. The situation of the sample members of WDCS under study reveals the following information.

Data presented reveal that 54.54 % of the total sample supported the statement regarding regular and timely availability of AI services against 45.46 % who did not. Again making use of breeding bulls is reported by 22.72% against 77.28 % who did not. Adequacy of breeding bulls to meet the requirement was mentioned by 6.36% against

93.64%.who did not avail. However, maintenance of breeding bulls by the society was available to only 12.72% as against 87.28%. The result as a whole provides a discouraging picture about breeding facility of WDCS. Unless and until breeding facility is provided at WDCS level the success of white revolution will not be achieved. An intervention in this respect needs no emphasis.

7. Credit facility at WDCS

Maintenance of dairy unit is cost intensive. Unless otherwise credit provision is made, the poor house hold cannot afford to it as the percentage of population in Odisha under poverty line is 48%. The need for credit facility is well recognized. While examining the availability of credit facility of WDCS, the following results are obtained.

Of the total sample 40% mentioned to have easy access to credit facility, 13.63% mentioned the rate of interest is higher while 66.33 % viewed credit facility are complicated to avail. On the whole it may be said that the sample members is not satisfied with credit facility.

8. Training and Extension

The State Government has made provision to provide technological back-up and capacity building facility for dairy entrepreneurs of the state. But observations revealed that many of the members of WDCS failed to avail training and extension facilities of the state government. The study made an attempt to find out real situation in the field and obtained the following information.

The percentage analysis of data revealed that 67.27% are happy with training programmes as those are need based, 65.45% mentioned extension programmes as per needs of the participants and equal percentage of sample mentioned the participation of members of WDCS in extension programmes. About 63.63% expressed satisfaction on extension service and 60% mentioned there is regular dairy extension programme, while 42.72% admitted training programme was regular. Comparing both districts, the respondents of Cuttack district avail more extension and training facility than Dhenkanal district. About 60 % of the sample appreciated extension programme while remaining 40 % offered opposite view.

9. Over-all satisfaction

It was under the frame work of the study to assess the overall satisfaction of the sample on different dimensions like sale of milk, price structure, veterinary services, vaccination, training programme and AI services. The overall pictures about satisfaction of members are presented below.

Table 5.4 : Overall satisfactions of WDCS members

Statements	Cuttack		Dhenkanal		Total	
	f	%	f	%	f	%
Satisfaction in milk sale	48.00	96.00	52.00	86.66	100	90.90
Satisfaction on quality and price of milk	45.00	90.00	54.00	90.00	99.00	90.00
Satisfaction of members to Veterinary Services	24.00	48.00	39.00	65.00	63.00	57.27
Satisfaction on quality of vaccine	39.00	78.00	48.00	80.00	87.00	79.09
Satisfaction on training programme	33.00	66.00	30.00	50.00	63.00	57.27
Satisfaction on AI Service	24.00	48.00	46.00	76.66	70.00	63.63
Average	-	71.00	-	74.72	-	73.02

On account of disposal of milk, the entire sample is satisfied (90.90 %) and on price of quality of milk about 90% were satisfied. On account of AI services, training programme, the satisfaction were expressed by 63.63% and similarly for veterinary service and vaccination by 57.27%. On the whole, majority of the members were satisfied on key performance areas which are strong determinants of success in co-operative sector.

The key performances of nine dimensions are presented in pictorial form to provide insight on importance of the dairy activities.

Table 5.5 : Ranking of key performance areas of WDCS

Key performance areas	Percentage	Rank
Milk procurement	82.11	1st
Overall satisfaction	71.96	2nd
Vaccination	64.17	3rd
Extension/Training programme	60.25	4th
Sale of milk & payment	56.05	5th
Cattle feed & Fodder	55.06	6th
Veterinary Services	54.07	7th
Credit Facilities	39.99	8th
Breeding	25.45	9th

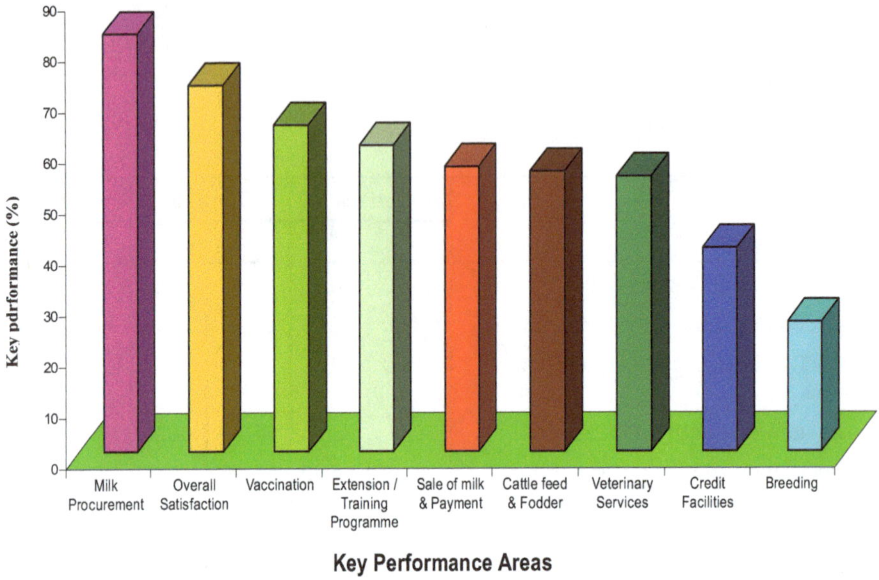

Fig. 5.4 : Ranking of key performance areas of WDCS

Out of nine key performance areas the minimum performance is found in case of breeding facility next to credit and veterinary services. The highest performing activities are found with milk procurement, followed by overall satisfaction, vaccination, extension training and disposal of milk and payment.

IV. Determination of Dairy Management Practices and Milk Production Profile of WDCS Members.

A. Dairy management practices

The research in veterinary science recommends to adopt better management practice for ensuring sustainable milk production level. Animal husbandry for success depends upon the adoption of practices recommended by scientists. The management practices cover cow shed, its sanitation, feed and fodder, breeding of animals, AI service, vaccination, veterinary and different other activities related to dairy animals.

The present study made an in-depth analysis of the dairy management practices and milk production profile of the area under study.

1. Cow shed

Cattle shed are very important from management point of view of dairy animals. Its situation in residential unit, structure, ventilation, space adequacy and overall cleanliness contribute towards improved management practices.

A look at the data reveal that 62.72% continue to keep animals in kaccha house and 26.36% under semi pucca houses whereas about 10.92% have opportunity to pucca cattle sheds. Comparatively participants of Cuttack district have better structured animal shed than the participants of Dhenkanal district as revealed through percentage analysis shown in Table 6.35.

2. Situation of cow shed

It goes without saying that animal requires enough space for ventilation, cleanliness and well location for sound health. While examining the situation of cow sheds of the sample under study the following results were obtained;

The finding reveals that 56.36% located their animal sheds within the residential complex while 43.64% had animal shed out side of their residence. Adequate space for the animals is expressed by 55.45% against 44.55% who expressed space as inadequate for the animals.

About 72.72% expressed that ventilation of their cow sheds are satisfactory against 27.28% of participants who expressed as not well ventilated.

Deciding the location of cow shed as deem to be suitable, 60.91 % realized to be suitable against 39.09% who did not agree.

About cleanliness of cow shed, 71.82% participants expressed as satisfactory against 28.18% who expressed as not satisfactory. Taking situation, space, ventilation, location, and cleanliness of the sheds, it is inferred that over all picture is encouraging except sheds inside the residential complex (56.36%).

However, the sample of Dhenkanal district were found to be in better position than the participants of Cuttack district with respect to situation of animal sheds, space, ventilation and cleanliness.

3. Fodder cultivation

Growing of green fodder has been emphasized for dairy animals invariably in all situations of Odisha. But unfortunately fodder cultivation is found to be given least importance. The investigation reveals that only 14 (12.72%) of the sample reported taking up fodder cultivation that too in a haphazard way against 96 (87.28%) who did not take up fodder cultivation.

The findings strongly recommends for special drive to grow green fodder by the members of WDCS.

4. *Feeds of dairy cows*

At present almost all dairy animals are fed with concentrates of different type of compositions. To rear dairy animals and run milk co-operative society, it is one of the major components in deciding the level of milk production. The study attempted to find out different type of feeds fed to dairy animals in different WDCS.

A look at the analysed data reveals that, about 53.63% of the sample fed the animals with concentrate, green fodder and roughages, 41.81% concentrate and green fodder, and same percentage dry roughage only. It is observed that about 22.72% fed their animals with green fodder whereas 16.36% depend upon only concentrate. On the whole it is revealed that roughage and green fodder find a place in most possible combinations of feeds fed for the animals. The earlier findings although does not reveal cultivation of green fodders, it is obvious that participants collect green fodder fed the animal. However, the name of fodder, area coverage and stage of growth of the green fodder fed to the animals are of no specific identification.

5. *Sources of cattle feed*

The cattle feed of different manufacturers and companies are available at village level. Equally some people also collect green fodder and roughages on payment. In finding out sources of cattle feed of the sample under study, the following results are obtained.

Finding reveals that 85.45% of sample collect their cattle feed from WDCS while 67.27% depend upon local traders and 49.09% with other sources. It may be mentioned here that during investigation it was found that WDCS deal with only concentrate where as for rice and wheat bran the sample have to depend upon local traders and other sources. The WDCS are not able to meet all feed requirements of dairy animals.

6. *Breeding practices followed by WDCS members*

For up gradation of local animals and maintenance of purity for selecting breeds, recommendation are given to make use of AI but on contrary at the village level mating with improved bulls and let loose for mating with stray bulls are not uncommon.

The study attempts to find out breeding practices adopted by the sample on 3 point scale varying from always, sometimes, never which yielded the following results.

Table 5.6 : Breeding Practices followed by WDCS Members

Practices	Always		Sometimes		Never		Total	
	f	%	f	%	f	%	f	%
Mating with improved bulls	22.00	20.00	26.00	23.63	62.00	56.37	110	100
Artificial incemination	70.00	63.63	31.00	28.18	9.00	8.19	110	100
Mating with stay bulls	12.00	10.90	23.00	29.90	75.00	68.20	110	100

The findings of the table reveal that about 20% of the sample always allow mating with improved bull, 63.63% take advantage of A.I., while 10.90% adopt late loose mating with stray bulls. On the contrary, 56.37% never adopted practices of mating with improved bull, 8.19% A.I. and 68.20% late loose mating with stray bulls. The picture provides an encouraging facts that the members of the majority WDCS adopt AI method for breeding of their animals.

7. Payment of AI services

AI services are provided by the veterinary department of State Govt. with more emphasis for milk co-operative societies. During investigation it was observed that the farmers pay Rs.10.00 to Rs. 30.00 for AI service at different localities of the surveyed area without any specified norm. However, the payment does not appear to be cost intensive to avail AI at door-steps.

8. Vaccination

The vaccination of animals is an important and integral part of dairy practices. Invariably all veterinary doctors recommend to vaccinate the animals for prevention of diseases, sound health and better milk production. But it is observed that some dairy owners are not particular of the importance of vaccination. While finding out the sources of vaccines and the use, the following information was recorded irrespective of the district under study.

As revealed from data, Govt. is the major source of vaccines followed by WDCS. No other agency in the area of the study deals with vaccines.

So far use of vaccine is concerned 58.18% of respondents reported that they regularly vaccinate their animals, 31.81% occasionally and 10.01% never adopt practice. Of the sample, depending upon WDCS, 29.09% do take vaccination regularly, 61.81% occasionally and 9.09% never adopt the practice. On the whole it reveals that about 10% of the samples do not vaccinate their animals. This needs intervention by the concerned agencies to motivate all the dairy owners to follow vaccination regularly.

9. *Cost of veterinary services*

At rural village of Odisha veterinary services are available against nominal payment fixed by the Govt. But there is no uniformity for door-step service and it differs from place to place depending upon the demand. While scaling the opinion of the respondents as very costly, costly, and not costly of the veterinary services availed by them, following results are obtained.

The respondents under study expressed that for major treatment of animals, the amount they pay becomes very costly for them (68.18 %), whereas 18.18 % feel costly as against 13.63% who expressed as not costly. In case of first aid treatment 22.72 % expressed to be very costly, 54.54 % costly and 28.73% not costly. On the whole, the major treatment of the animals was found to be costly, where as taking of vaccination or for minor first aid was not costly.

B. Profile of milk production

Milk production in Odisha is reported as 1651.33 thousand MT with productivity level of 166 kg/100 days, 1167 kg/300 days for Deshi, crossbreed cows respectively. The measure objective of WDCS is to increase production level of milk and making supply to OMFED. The objective is related to production, consumption and disposal of milk to OMFED. The WDCS are meant for maintenance of dairy animals and production of milk. The milk is to be supplied to OMFED which is operating in Anand pattern. The present study made an indepth analysis of milk production profile covering the aspects like possession of different breeds, production of milk, consumption and disposal of milk to cooperative society.

Possession of breeds

The farmers of Odisha are in habit of rearing deshi cows since time immemorial. Presently the cross breed have added to the stock besides keeping buffaloes. The present study investigated possession of desi, cross-breed and buffalo in the area of study.

1. Possession of desi cow

Desi cows are very popular among the farmers of Odisha for the reasons of low investment, easy care, easy milking and preference of consumers.

Of the sample 47.27% of the respondents possess 1-3 nos desi cow, 10.90% possess 4-6 nos desi cows and 2.73% possess as much as 7-9 cows. It is observed that one of the respondents possess more than 10 cows. Of the sample as much as 39.09 % did not possess any desi cow. Comparing both the districts, more number of desi cows are found to be with the members of Dhenkanal district.

2. Cross breed cows

According the census 2001 there are about .74 million (5.30 %) numbers of cross breed cows in the state comprising of mostly Jersey, Holstein Frisian, red sindhi. While investigating the sample respondents, the possession of cross breeds were found as given table below;

Analysis of the data reveal that as high as 55.45% possess 1-3 numbers of cross breed cows, 8.18% possess 4-6 numbers of cross breed while 3.63% possess 7-9 cows. It is interesting to note that 32.74% did not posses cross breed cows at all. Comparing the two districts more no. of cross breeds are found in the district of Cuttack.

3. Possession of buffalo

The population of buffalo in the state as per census 2001 is reported to be13.88lakh(5.9%). The buffaloes of Odisha mostly comprised of Parlakhemundi, Sambalpuri, Chilika and Manda which are region specific. There are some locations of Odisha where buffaloes are reared and in some locality the population is very low. In finding out status of buffaloes with sample respondents the following results were obtained.

The data reveal that about 90 % of the respondents did not possess buffalo at all, while 5.45% had 4-6 numbers buffaloes, 2.72% had 7-9 numbers buffaloes, 1.82% having 1-3 buffaloes. It is interesting to note that only one respondent in Cuttack district posses buffalo against 10 members of Dhenkanal district. On the whole the possession of buffalo is found to be very negligible in the coastal tract of Odisha.

4. Milk production pattern of milch animals of WDCS members

To find out a clear picture of the situation, attempt was made to ascertain the milk production status of different breeds maintained by the members of WDCS. The findings of the study are presented in the table below.

Table 5.7 : Comparison of Milk production Among Deshi, Cross-breed & Buffalo

Litres/ day	Score	Desi cow				Cross breed				Bufalo			
		Cuttack	Dhenkanal	Total		Cuttack	Dhenkanal	Total		Cuttack	Dhenkanal	Total	
		f	f	f	%	f	f	f	%	f	f	f	%
1-3	1	12	28	40	36.36	10	24	34	30.90	0	7	7	36.36
4-6	2	5	3	8	7.27	10	5	15	13.63	0	2	2	1.81
7 & more	3	2	0	2	1.81	8	2	10	9.09	1	3	4	3.63
No milk	0	31	29	60	54.56	22	29	51	46.38	49	48	97	88.20
Total		50	60	110	00	50	60	110	100	50	60	110	100

As revealed in the table above, in case of Deshi cow 36.36% produce milk within limit of 1-3 lit./day, 7.27 % produce 4-6 lits and 1.8 % produce more than 7 liters/day. On the whole about 54.56 do not possess Deshi cow.

Cross breeds are high yielders. The results reveal that about 30.90% produce 1-3 liters milk/day while 13.63% produce 4-6 liter/day while 9.09% produce more than 7 litres per day. However, 46.38% did not possess cross breed cows.

In case of buffaloes, results revealed that very few members possess them that too in Dhenkanal district. As much as 88.20% do not possess buffalo at all where as 6.36% produce 1-3 liters, 1.81% produce 4-6 liters and 3.63 % more than 7 liters per day. The picture clearly shows that cross-breed cows comparatively produce much more milk than deshi cows and buffaloes.

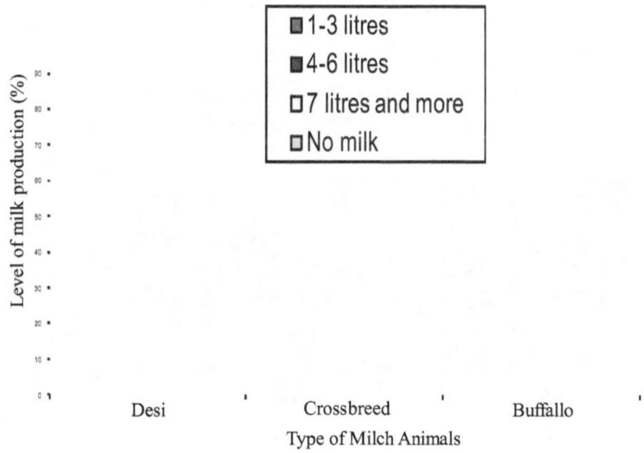

Fig. 5.5 : Milk production pattern of WDCS members in two districts

5. Milk consumption pattern of WDCS members

Milk consumption in the household is an indicative of taking care of health. In order to assess the situation of the consumption of milk obtained from deshi, cross-breed and buffalo, the information collected, analysed as revealed in the table below.

Table 5.8 : Comparison of consumption pattern milk obtained from deshi, cross breeds and buffalos

Litres sold per day	Score	Desi cow				Cross breed				Bufalo			
		Cuttack	Dhenkanal	Total		Cuttack	Dhenkanal	Total		Cuttack	Dhenkanal	Total	
		f	f	f	%	f	f	f	%	f	f	f	%
<5	1	19	25	44	40.0	11	24	35	31.81	0	6	6	5.45
>5	2	0	6	6	5.45	17	7	24	21.81	1	6	7	6.36
No milk	0	31	29	60	54.54	22	29	51	46.38	49	48	97	88.19
Total		50	60	110	100	50	60	110	100	50	60	110	100

The above table revealed that comparing the consumption of milk obtained from desi cows, cross breed cows and buffalos, 13.63 % of the respondents consume cow milk less than 0.5 lit followed by 5.45 %of cross breed cow milk. However, people do not consume buffalo milk as evidenced from the table. Maximum respondents (28.18 %) were consuming milk of cross breed cows within the range from 0.6-1.0 litre as against 16.36 % in case of desi and 1.81 % in case of buffalo. Equal percentage of respondents i.e., 7.27 % consumed the milk of desi and crossbreed cows more than 1 litre milk/day.

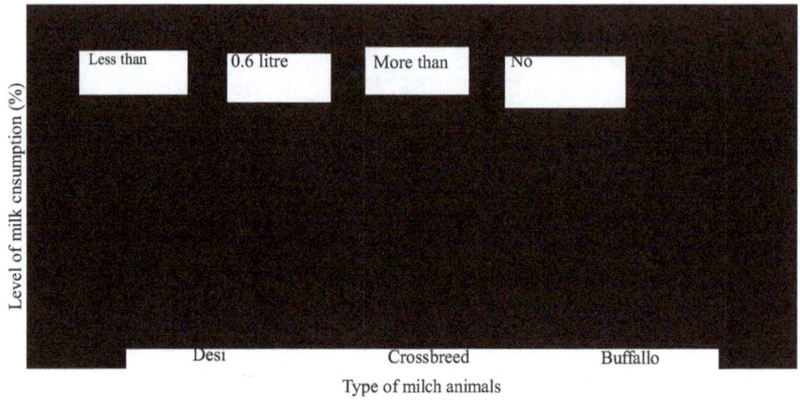

Fig. 5.6 : Milk consumption pattern of WDCS members in two districts

6. *Milk sale pattern of WDCS members*

The WDCS are bound by norms to supply milk to the society. An examination of situation regarding disposal of milk to society reveal the following information as depicted in the table below.

The owner of Deshi cow (40%) supply less that 5 liters milk to society against 5.45% more than 5 liters of milk.

In case of cross breed 31.81% dispose up to 5 liters/day to society against 21.81% where dispose more than 5 liters of milk.

In case of buffalo 5.45% supply 5 liters milk/day to society against 6.36 % of milk more than 5 liters per day.

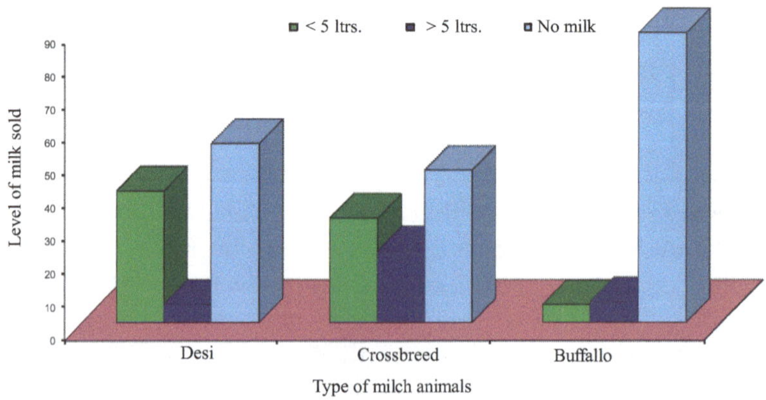

Fig.5.7 : Milk sale pattern of WDCS members in two districts

7. Milk production, sale and consumption profile

Comparing the production, sale and consumption of the milk of the participants, the following informations were received;

Table 5.9 : Comparative milk production, sale & consumption profile (score analysis)

Type of animal	Production	Sale	Consumption
Deshi	0.56	0.50	0.68
Crossbreed	0.85	0.75	0.83
Buffalo	0.18	0.18	0.03

The score analsys presented in the table above reveals that members possessing cross breeds are much ahead in production, sale and consumption at house hold level over deshi cows. The buffalos owners under the sample size are invariable low in production, sale and consumption.

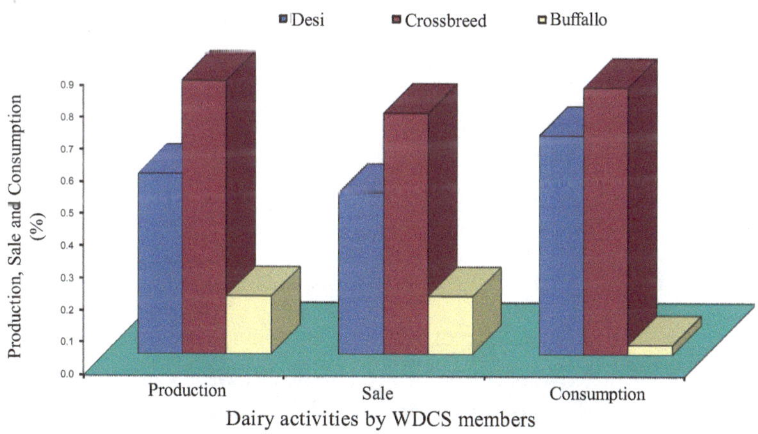

Fig. 5.8 : Comparative milk production , sale and consumption profile (Score analysis)

V. Extent of Participation of Women Respondents in Dairy Activities

Women perform many activities which have not recognized as expected. They are called invisible workers. Compared to agriculture, horticulture, and fishery women take more responsibility in animal husbandry activities. While examining the participation of women in different activities relating to dairy management, the informations were received as contained below. The activities have been categorized into 5 different groups like arranging feed and fodder, breeding practices, treatment of animal, cleaning and sanitation, milking and milk marketing.

1. Feed and fodder

The feed and fodder consisted of arranging feed and fodder, concentrates feeding and watering of animals. The participation of women in watering of animals was found to be the highest (77.27%) followed by feeding (70.00%), arranging concentrate (35.45%) and arrangement of feeds and fodder (23.63 %).

Comparing with self, their counter parts participated more i.e., 56.36 % in arranging feeds and fodder, 51.81 % in arranging concentrates. The service of the servants was found to be negligible in all cases. With husband their participation was highest in arranging feeds and fodder, feeding of animal and watering in order.

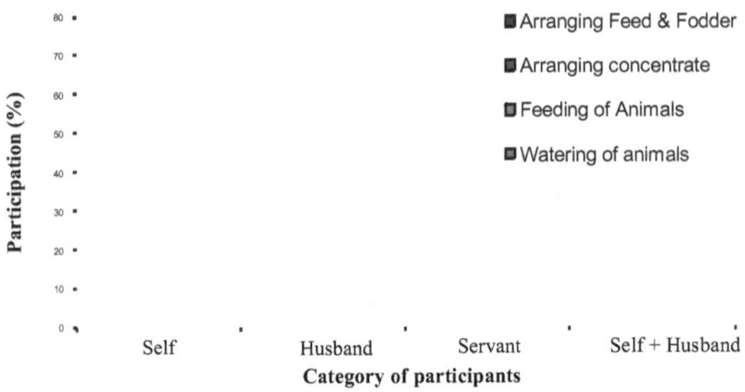

Fig. 5.9 : Participation of women in arranging feed and fodder

A look at findings reveal that the level of participation of self (women), husband, servant, self and husband were same except arranging feed and fodder. The participation of husband in Dhenkanal district was more than in Cuttack district.

2. Breeding of dairy cows

Breeding of dairy cows consisted of the activities like taking animals for mating which is mostly performed by the husband (81.81%) while only 16.36% of their counter parts did this job.

The study reveals that comparatively husbands are involved more in breeding practices in Cuttack district than Dhenkanal district.

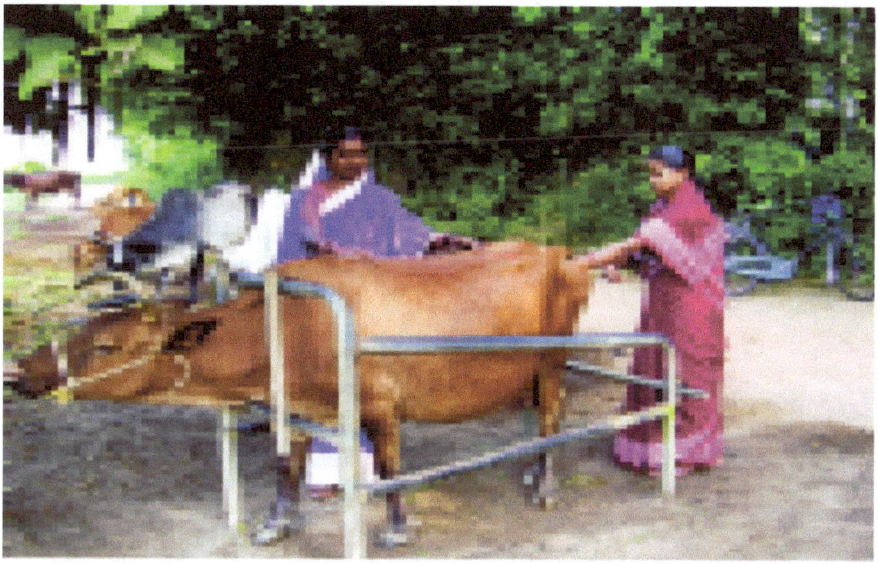

Plate 1 : Women doing artificial insemination

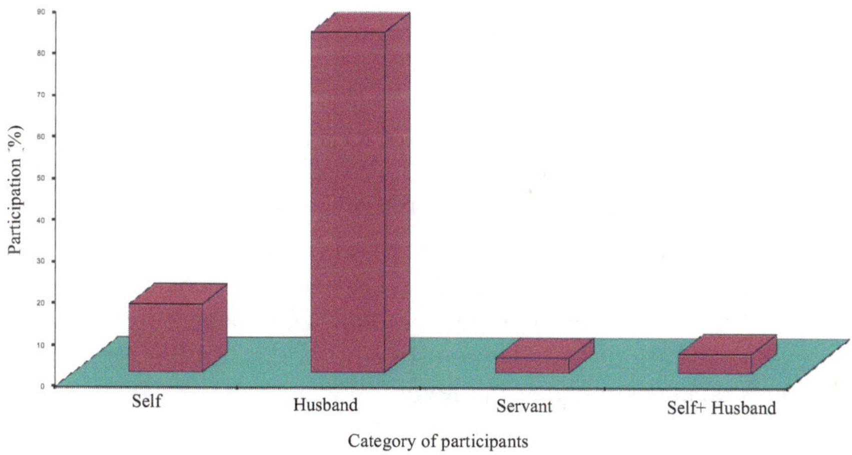

Fig. 5.10 : Participation of women in breeding practices

3. Treatment of dairy cows

The treatment of animals was studied in two aspects i.e. purchase of medicines and taking animals to the veterinary hospital for treatment. In both the cases, women alone found to be less involved than their counter parts. However, with husband, their participation is reported up to 11.81% for purchase of medicine and 6.36% taking animals for treatment.

The data revealed that husbands were more involved in both the treatment activities such as taking animals for treatment (71.81 %) and purchase of medicines (66.36 %).

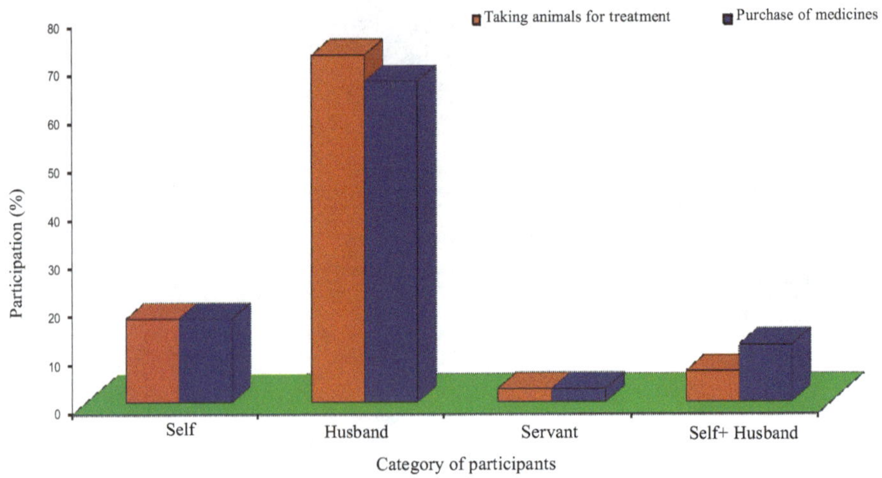

Fig. 5.11 : Participation of women in treatment of dairy animals

Comparing two districts the involvement of husband is found more in purchase of medicines in case of Dhenkanal districts than their counterparts. It is found that involvement of members were more in both the cases such as taking animals for treatment and purchase of medicine in Dhenkanal districts than the Cuttack district.

As seen from the analysed data that involvement of the members in taking animals for treatment and purchase of medicines in Dhenkanal district were much better than the members of Cuttackdistrict. However, the reverse in case of involvement of husband participation, servant, self and husband were at equal level.

4. Cleaning and sanitation

The cleaning and sanitation consist of the components like collection of dung from shed, grooming the animals, taking buffaloes for wallowing.

It is found that women participation in grooming the animals is 74.54% as against 15.45% done by husband and 8.18% done by both husband and self. So, it is understood that women's participation is much higher in caring the animals. In case of collection of dung from the sheds, keeping clean of the shed women's participation is much higher (83.63 %) than their husband, servant and both servant and husband respectively. However, taking buffaloes for wallowing husband's participation is much higher than others. In case of cleaning of shed and grooming of animals, women participation was found much higher than their counter parts. It showed that cleaning and sanitation of animals shed are mostly taken up by the housewives.

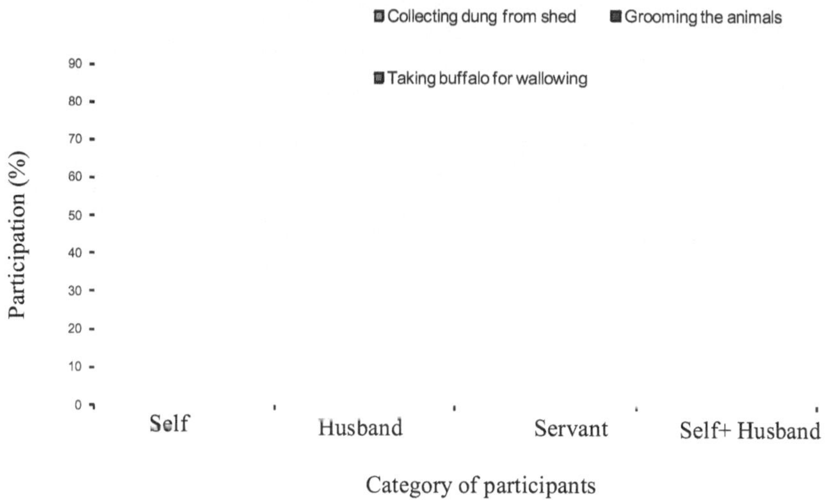

Fig. 5.12 : Involvement of women in cleaning and sanitation of cowshed

Comparing two districts under study as revealed in the table above participation of women in cleaning and sanitation activities in both the districts at the same level with respect to collecting dung from shed where as participation of women was more in case of Dhenkanal district as compared the Cuttack district with respect to grooming the animals. However, the table revealed that participation of the self, husband and self is very negligible or nil in taking buffalo for wallowing in Cuttack district than Dhenkanal district. So, it is because that the members of the Cuttack district prefer to keep more crossbreed dairy cows than buffaloes compared to Dhenkanal district.

5. Milking and milk marketing

Milking and milk marketing was studied taking into account of milking of cows, carying milk to cooperative societies, washing of the milk trough, decision to sell and purchase of cow and collection of money. The result reveals that the women take leading role in carrying milk to cooperative societies (55.45 %), washing the milk trough (85.45 %) and collection of money (55.45 %) and milking (50.0 %) , where as their counter parts found to be more involved in case of selling and purchasing of animals. However, about 52.72% of sample expressed that they decide with their husband in purchase and sell of cows. On the whole it seems that women take leading role in milking and marketing.

Plate 2 : Women milking the cow and pouring milk in WDCS

Fig. 5.13 : Participation of women in milking and marketing

In summarizing the findings, the independent role of women for the five activities are presented in a pictorial form.

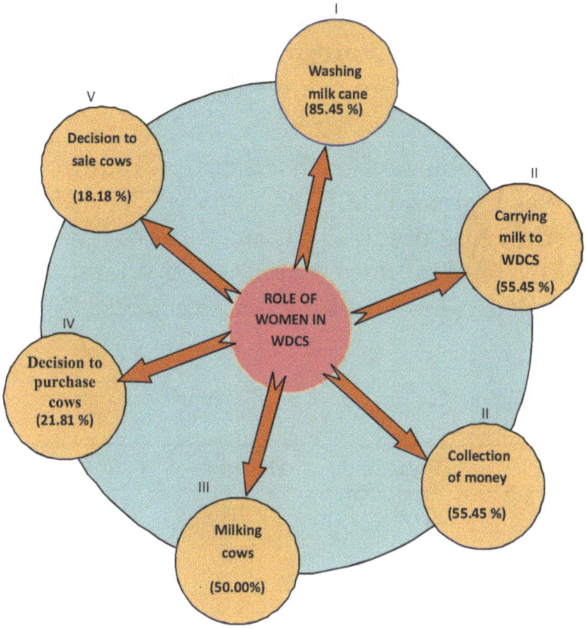

Fig. 5.14 : Independent participation of women in milking and milk marketing activities

6. District-wise participation of women

Further, an attempt was made to find out difference if any in the participation of women in dairy management practices between the district of Cuttack and Dhenkanal .

Analysis reveals that participation of women in dairy management system as obtained under self, husband, servant, self + husband was maximum being observed in case of self, followed by husband, self and husband. In other wards, the participation of women in dairy management practices was higher incase of Dhenkanal district compared to Cuttack district.

VI. Entrepreneurial Characteristics of Women Dairy Cooperative Society Members.

WDCS is an entrepreneurial unit of livelihood. Many families in rural areas depend upon dairy animals for their living. Milk production and productivity and the remunerative price are the key factors for the successful dairy units. It is hypothesized that entrepreneurial characteristics of the individuals are related with success or failure of the enterprise. Within the frame work of the study an attempt was made to identify entrepreneurial characteristics of the WDCS members on ten important aspects such as Quality consciousness, Desire of achievement, Optimistic, Foresightedness, Independent, Future planning, Good organizer, Innovativeness, Information seeking, hard work.

A glance at the table reveals that the gaps in different aspects vary from 27.33% to as high as 62.00%. The analysis reveals that maximum gap is observed in case of foresightedness (62 %) followed by quality consciousness (58.66 %), future planning (55 %), information seeking habits (53 %), independent in taking decision (53.33 %) and hard work (50.66%).

Table 5.10 : Entrepreneurial behaviour of WDCS members

Characteristics	Cuttack	Dhenkanal	Average	Gap (%)
Hard work	1.84	1.18	1.51	50.66
Desire of achievement	2.04	1.57	1.80	40.00
Optimistic	2.34	2.03	2.18	27.33
Foresightedness	1.70	1.11	1.40	62.00
Independent	1.43	1.38	1.40	53.33
Future planning	1.45	1.26	1.35	55.00
Good organizer	1.60	1.56	1.58	47.33
Innovativeness	1.70	1.61	1.65	45.00
Information seeking	1.60	1.23	1.41	53.00
Quality consciousness	1.45	1.04	1.24	58.66
Average	1.71	1.39	1.55	49.73

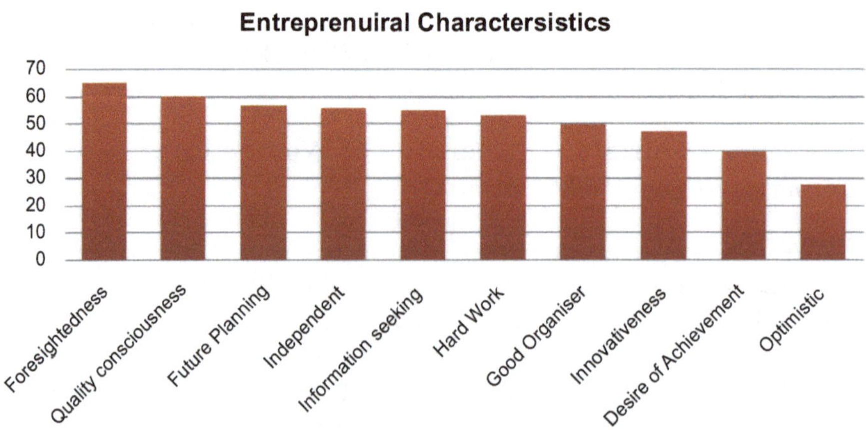

Fig. 5.15 : Gap in entrepreneurial bebaviour of WDCS members

The other aspects in which the gap is observed are to be good organizer, innovativeness, desire for achievement and optimistic in maintenance of enterprise. Taking into consideration of Cuttack and Dhenkanal district, the gap is observed to be more in case of Dhenkanal district than Cuttack. The overall gap is 49.33%, and district wise gap is found to be 43% and 53.66% for Cuttack and Dhenkanal district respectively. So the overall conclusion is that the members of WDCS have not been educated or trained to exhibit their entrepreneurial characteristics. The high percentage of gap indicates need for intervention in entrepreneurial aspects to boost up their production and productivity level leading to success of the society. The difference between Cuttack and Dhenkanal districts is due to higher level of education, closeness to milk processing unit, better socio-economic status and exposure to mass media.

VII. Impact of WDCS on Socio-economic Status of the Members

In sociological studies impact assessment of any programme receives for analysis and interpretation. The impact is normally defined as the difference between the situation before and after implementation of programme. Within the framework of the study, an attempt was made to find out impact of WDCS on five major dimensions like possession of dairy animals, material possession livelihood status, investment pattern and capacity building. Change in these aspects were studied to ascertain the direct and indirect impact of WDCS to which the sample members belong.

1. Change in possession of dairy animals

It is hypothesized that with the growth of WDCS, there would be increased number of dairy animals along with total milk production and milk consumption rate. The rate of milk production is also is the product of the change over time. To ascertain the change in possession of dairy animals under production level, the samples were interviewed to reveal their pre and post possession of dairy animals and status of milk production, consumption activities.

A look at the table reveals that highest change has been observed in case of rate of milk/lit (77.27 %) followed by milk production per day (74.54 %) and possession of crossbred cow (63.63 %) and milk consumption/day (48.18 %). In other words possession of buffalo and desi cow showed very slow rate of change over the years among the members of WDCS. It is therefore, confirmed that establishment of

WDCS has significant impact on the economic life of the members but the society did not have any significant impact on the members to increase buffalo population.

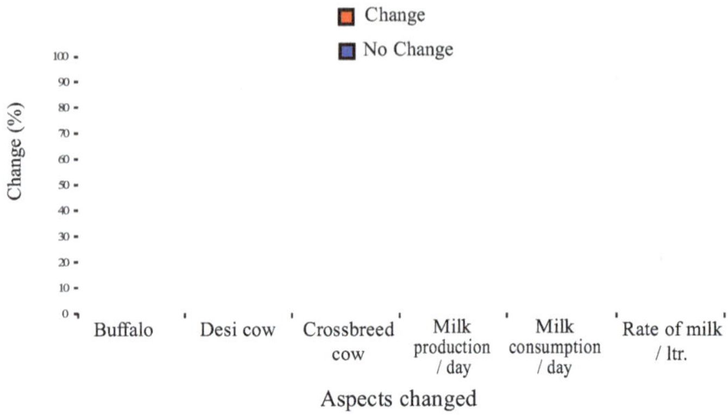

Fig. 5.16 : Change in position of dairy animals and milk production

2. Change in material possession

Material possessions are the good indications to reveal change in status. Keeping all other factors constant, efforts were made to assess the impact of WDCS on the material possession of the respondents under study. The material possession included bicycle tractor, pump set, T.V., VCR and change in area under fodder production.

Table 5.11 : Change in material possession

Materials	Change		No change		Total
	F	%	F	%	f
Cycle	47	42.72	63	57.28	110
Motor cycle	19	17.27	91	82.73	110
Water pump	7	6.36	103	93.64	110
Television	36	32.72	74	67.28	110

Analysis table reveals that significant change has occurred in possession of cycle, T.V, automobile, area under fodder cultivation and possession of water pump in order. However, the change seemed to be slow for the reason of investment and the profit perceived by the sample under study.

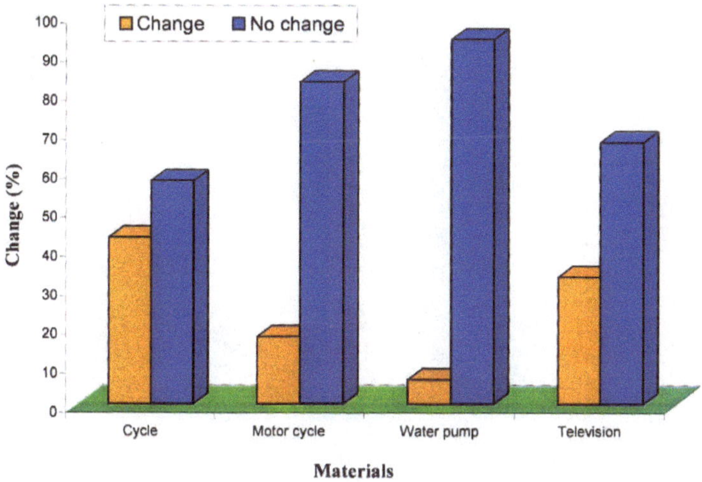

Fig. 5.17 : Change in material position due to WDCS

3. Change in livelihood status

In attempting to find out change in life style of sample under study as much as seven parameters were taken into consideration. The change with respect to livelihood of the respondents under study reveal the following results;

A look at the table reveals that highest changes have been experienced by the sample in education of the children, followed by annual income, better health care, contact with Animal Husbandry extension functionary, access to the media, building of the house and purchase of the land in order. In other wards, the priority area of the change is found to be over children education, annual income, health care and extension contact.

Table 5.12 : Change in livelihood status

Statements	Change		No change		Total
	f	%	f	%	f
Annual income	90	81.81	20	18.19	110
Health care	86	78.18	24	21.82	110
Purchase of land	5	4.54	105	95.46	110
House purchased or build	25	22.72	85	77.28	110
Children education	92	83.63	18	16.37	110
More access to mass media	69	62.72	41	37.28	110
More contact with AH extension functionary	76	69.09	34	30.91	110

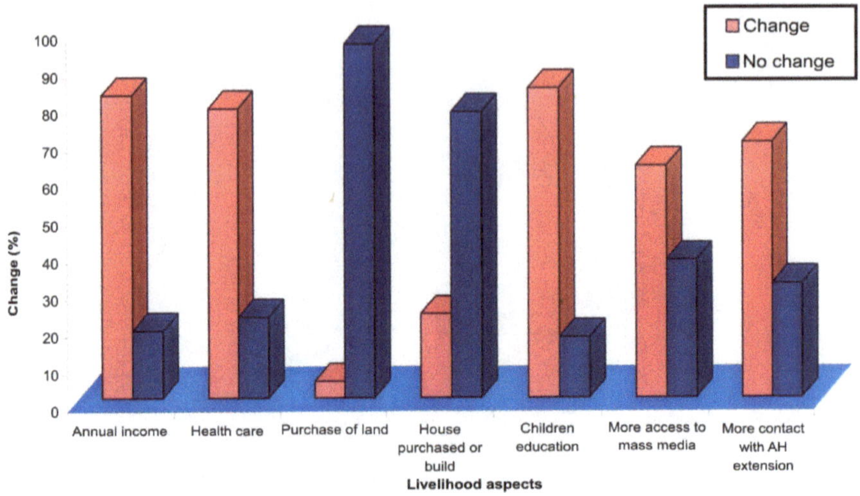

Fig. 5.18 : Change in livelihood status of WDCS members

4. Change in pattern of investment

With better-assumed income, people normally decide the area of investment. Keeping the occupation in view, the investment pattern of the sample was studied against some specific parameters which yielded following results.

Table 5.13 : Change in investment pattern

Investment pattern	Change		No change		Total
	f	%	f	%	f
Spend in the business	30	27.27	80	72.28	110
Make saving in bank	43	39.09	67	60.91	110
Spend for purchase of land	10	9.09	100	90.91	110
Spent purchase construction of house	34	30.90	76	69.10	110
For better health care	89	80.90	21	19.10	110
Better observation of social functions like marriage of sons/daughter	25	22.72	85	77.28	110
Mass media	63	57.27	47	42.73	110
Spend for education of children	84	76.36	27	23.64	110
Extension contracts	42	38.18	68	61.82	110

The pattern of investment as indicated above reveals primarily for better health care, education of children, contacting mass media, saving in bank and extension contact. However, minimum priority is found with respect to investment on purchase of land, better observation of social functions and expenditure in other. The members of the society are involved in dairy and hardly prefer other avenues of life. However, considering the investment patterns they mostly invest on health care and children education.

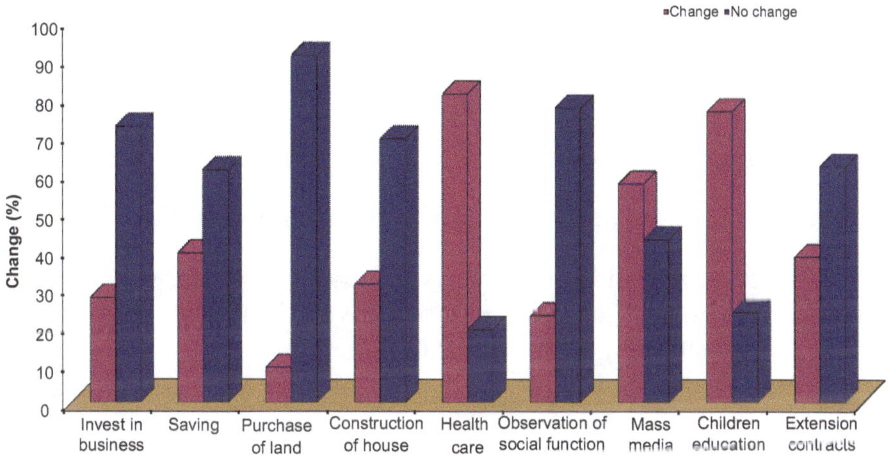

Fig. 5.19 : Change in investment pattern of WDCS members

5. Change in capacity building

The Government as well as the NGOs are organizing a number of programmes to enrich the skill of the rural women. The sample under study in the matter of capacity building through different programmes were found.

A look at the data reveal that the women under study have more participation in village meetings (75.45 %), followed by training (66.36 %), demonstration (38.18 %) and krishi mela (15.45 %). However, their participation in field days and exhibition is found to be limited. Overall participation in the capacity building programme was found to be at increasing rate.

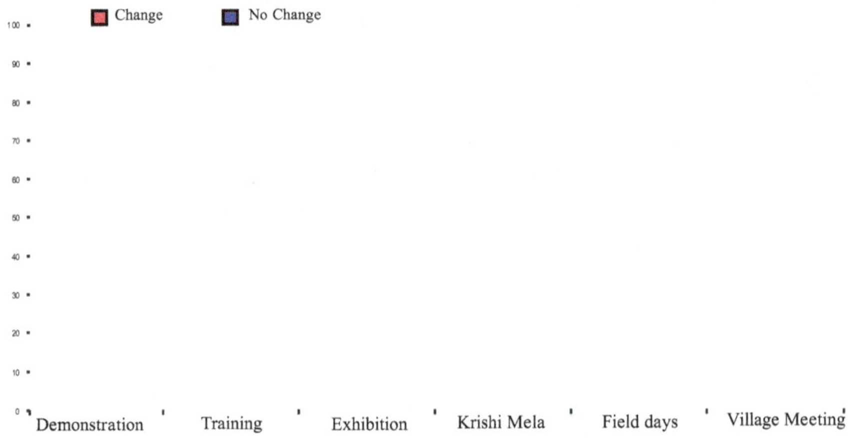

Fig. 5.20 : Change in capacity building activities of WDCS members

6. Overall changes in different dimensions due to WDCS

Change in five different aspects like possession of dairy animals, material possession, livelihood status, investment pattern, capacity building, the following results were obtained.

Table 5.14 : Over all changes due to WDCS

Dimension of change	Average change (%)	Rank
Possession of dairy animals	47.87	II
Material possession	21.81	V
Livelihood status	57.52	I
Investment pattern	42.01	III
Capacity building	34.39	IV

A look at the table reveals that maximum change has occurred in livelihood status followed by possession of dairy animals, investment pattern, capacity building, possession of materials in order. The change varies from 21.81 % to as high as 57.52 % indicating a favourable change in the life style of member of WDCS.

VIII. Correlating Variables and their Interrelationship

For smooth functioning of cooperative societies, a number of variables play their roles either in single or combination. An attempt has been made to study the interaction of these variables affecting the performances of WDC. In the following descriptions efforts are made to find out the relationship among socio-personal variables with key performance areas (KPAs), socio-personal variables with entrepreneurial behaviour of the sample under study.

1. Socio-personal variables and key performance areas (KPAs)

Out of ten socio-personal variables as much as seven variables were selected on the basis of their relative importance to correlate with KPAs in relation to study. While examining correlation between socio-economic variables and KPAs, the following results are obtained.

Table 5.15 : Correlation between socio economic status and key performance areas.

Code	Key performance areas	'r' value
KPA_1	Milk procurement	0.03456
KPA_2	Sale of milk and payment	0.0375
KPA_3	Veterinary service	0.10603
KPA_4	Cattle feed and fodder	0.1630
KPA_5	Vaccination	0.0892
KPA_6	Breeding	0.1647
KPA_7	Credit facility	0.1037
KPA_8	Training and extension	0.2116
KPA_9	Satisfaction	0.0282

A reference to the table above may be mentioned here that socio-economic status was calculated on the score assigned to variables like age, respondent's education, spouse education, occupation, caste, family type, herd size and house structure. The total score obtained by individual respondent was summed up and classified into high, medium, low assigning 3, 2, 1 score respectively. The KPAs score were calculated on the basis of the total score obtained by the respondents taking into sub-component of each area.

A took at the findings reveal that training and contact with extension persons was found to be significant by associated with socio-economic status of the respondent under study. While all other areas (KPAs 1 to KPAs 9 except KPAs 8) were found to be non-significant. It is therefore, inferred that socio personal variables have little influence over milk production, sale of milk, veterinary service, arrangement of cattle feed, taking of vaccination breeding credit facility and overall satisfaction. It needs more interventions of technology and marketing.

2. Socio-personal variables with an entrepreneurial behaviour

Entrepreneurship is an innovative trait with individual which can be developed according to the situation. All the individuals do not possess these traits in similar fashion. It is hypothesized that entrepreneurial behaviour varies as per socio-personal variables. While examining the relationship between socio-personal factors and entrepreneurial behaviour the following results were obtained.

As contained in the table below the entrepreneurial behaviour of the sample have been calculated taking 10 important variables such as hard work, desire for achievement, optimism, foresightedness, independent thinking future planning, organizing capacity, innovativeness, information seeking habit and quality consciousness.

Table 5.16 : Socio-personal status and entrepreneurial behaviour

Socio-cconomic profile	'r' value
Age	0.346
Respondent education	0.740
Spouses education	0.552
Family type	0.006
Occupation	0.088
House structure	0.108
Caste	0.132

The selected socio personal variables include, age, respondent education, spouse education, family type, occupation, house structure and caste. The results reveal that age, education of the respondent and education of spouse are two most important contributing factors to develop entrepreneurial behaviour among the members. The caste appear to be significant while family type, occupation and house structure did not reveal to bear relationship with entrepreneurial behaviour. It is inferred that entrepreneurial behaviour has to be developed taking to correlating factors into account for achievement in enterprise.

3. Correlation matrix of KPAs

In order to gain comprehensive insight into the key performance areas of WDCS which are interrelated and interwoven , a matrix was constructed applying correlation which yielded the following results.

Table 5.17 : Correlation matrix

	KPA_1	KPA_2	KPA_3	KPA_4	KPA_5	KPA_6	KPA_7	KPA_8	KPA_9
KPA_1		0.061	0.052	0.140	0.104	0.029	0.075	0.050	0.142
KPA_2			0.157	0.222*	0.031	0.187	0.335**	0.046	0.243*
KPA_3				0.306**	0.472**	0.387**	0.180	0.356**	0.540**
KPA_4					0.187	0.236*	0.316**	0.450**	0.350**
KPA_5						0.128	0.118	0.327**	0.513**
KPA_6							0.256**	0.075	0.277**
KPA_7								0.255**	0.333**
KPA_8									0.491**
KPA_9									

Correlation matrix reveals that the milk production is correlated with sale of milk and trend of payment is related with feed and fodder activity, credit facility and overall satisfaction of WDCS members. Availability of veterinary service is related with all key performance areas with credit facility, while feed and fodder activity is associated with breeding of dairy animals, credit facility, training and extension programme and overall satisfaction. Vaccination programme is found to be associated with training and extension and overall satisfaction . Breeding of dairy cows is an important component of dairy programme. It is found to be associated with credit facility as well as overall satisfaction with members. Credit facility is found to be associated with training and overall satisfaction while training itself is associated with overall satisfaction. The overall findings reveal that overall satisfaction is an important factor which is found to be associated with all key performance areas except regularity of milk procurement process which may be a case of exception. In short the correlation matrix showing the interrelationship between different key performance areas lead to advice to strengthen the relationship through appropriate steps.

IX Constraints Faced by the Members of WDCS

Women Dairy Project (WDP) is operating in Odisha in 7 district of covering 758 no. of WDCS. The functioning of WDCS is not independent of problems directly or indirectly. Attempt was made to enlist the problems faced by WDCS in the area under study.

Table 5.18 : Constraints faced by the members of WDCS

Constraints	Score	Rank
Lack of credit facility	9.24	I
Non-availability of good breeds, feed and fodder	8.78	II
Competition in the market	4.75	VIII
Limited mobility as women member	8.63	III
Excess workload at home	7.84	IV
Low level of education	6.79	V
Gender bias	6.50	VI
Lack of risk bearing capacity	6.11	VII
Improper function of WDCS	3.78	IX
Lack of support from milk union	2.27	X

Looking at the content of the table it is clear that lack of credit facility and non-availability of good breeds, feed and fodder are the major problem followed by limited mobility of women and low level of education. The other factors that inhibit growth of WDCS are, gender biasness and lack of risk bearing capacity. However, competition in the market is found to be at 8th rank so far as the constraints are concerned. The other factors like improper function of WDCS and lack of support for union people and other miscellaneous problems come in order.

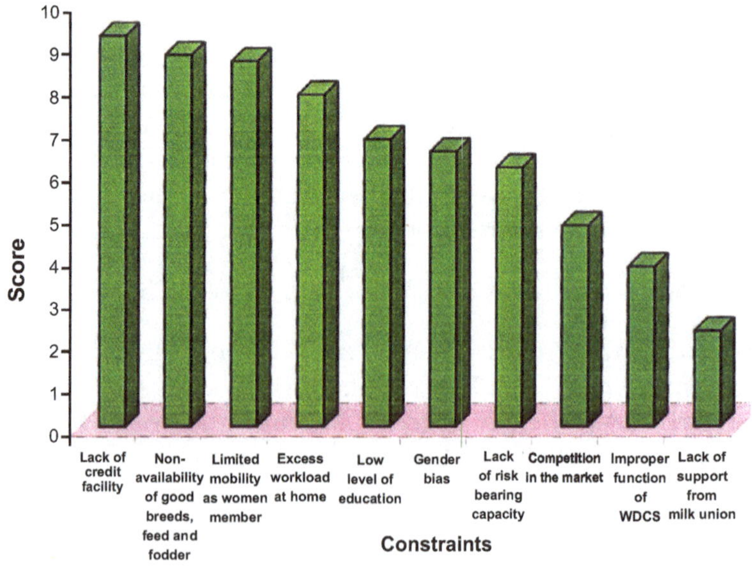

Fig. 5.21 : Constraints areas faced by the WDCS members

The results reveal that the members of WDCS encountered a variety of problems of different dimensions which need to be sorted out for smooth functioning of WDCS.

X. Suggestions given by WDCS Members for Improvement.

It has been seen that when people work in an organization or institution develop a sense of belongingness and mutual responsiveness. As a part of their responsibility they strive hard for development of the organization. In the subject under study, the members of WDCS by virtue of their participation in WDCS activities they gained some experience on the functioning of WDCS which are controlled under the by-law of the society. Further, society members from time to time depend upon the guidelines of the State Govt. and District Unions for their operations. However, in certain areas society members feel some deficiencies for which the societies are not able to perform to their expectation. An attempt has been made in this study to document the suggestions of the society members for further improvement of the society. The suggestions were asked in seven important areas such as procurement of milk preparation and sale of milk products, improving availability of cattle feed, to grow more green fodder, avail efficient veterinary services, to undertake regular extension programmes and above all to run WDCS efficiently.

1. Improvement of milk procurement in society

One of the important indicators for the performance of the society is the milk procurement by the society per day.The data are presented in the following table.

Table 5.19 : Improvement of milk procurement

Suggestions	Responses	
	f	%
Increase sale price of milk	110	100.00
Increase society members	68	61.81
Supply cows at subsidized rate	68	61.81
Honesty in milk fat measurement	89	80.90
Timely payment	82	74.54
Average	-	75.81

The table reveals that all the respondents (100%) are in opinion of increasing sale price of milk followed by 80.90 per cent respondents opined to maintain honest in milk fat measurement at society level and timely payment (74.54%), where as the members gave equal importance i.e. 61.81 per cent to both i.e. increasing society members and supply of cows at subsidized rates.

2. Preparation and sale of milk product

Non of the WDCS under study prepares milk product and sale them. When asked to start preparation and sale of milk product, the suggestions of the respondents were documented and detailed below.

The analysed data reveal that out of the total respondents, 61.81% respondents opined for imparting training on preparation of milk product followed by provision of storage facility (54.54%) and equal opinions i.e. 50.90 per cent were for supply of milk products to the society and bringing awareness among the public on consumptions of milk products.

3. Improvement of Availability of Cattle Feed

For success of any dairy unit, besides green fodder, supplementation of concentrate with cattle feed is important. However, it is seen that in most of the WDCS the members fed their cows home made cattle feed, grass and hay. When asked their suggestions, they replied with constructive views as below.

From the data it is inferred that 89.09 per cent of the respondent were in opinion of reducing cattle feed cost followed by increasing faith on quality of OMFED feed (77.27 %) and (53.63 %) opined to provide storage facility for cattle feed.

4. Improvement of availability of green fodder

The WDCS members were aware of the benefits of feeding green fodder to their cows. However, it was observed that in most of the WDCS there was scarcity of green fodder and farmers rarely fed their cows with green fodder . To improve the availability of green fodder in the area their suggestion were documented and analysed in the Table 5.20

Table 5.20 : Improvement of availability of green fodder

Suggestions	Responses	
	f	%
Provision of land for food cultivation	93	84.54
Irrigation facility	104	94.54
Subsidized fertilizer	76	69.09
Imparting training	83	75.45
Average	-	80.90

A glance at the table above reveals that maximum respondents i.e. (94.54%) were in opinion for improving irrigation facility in the society area where as 84.54% suggested to provide land by the Government to the society for fodder cultivation followed by imparting training (75.45%) and supply of fertilizer at subsidized rate (69.09 %).

5. Improvement of availability of veterinary services

The WDCS members need the services of veterinary doctors at different times to treat their cows. When enquired, most of the members expressed their dissatisfaction on non-availability of veterinary services at the time of need. Their suggestions for improving the veterinary services were documented and data after analysis revealed that 84.54 per cent of the members of the societies expressed to improve the communication facilities followed by equally expressed (65.45%) for making appointment of Veterinary doctor and provision of mobile van facility respectively.

6. Promotion of extension programmes

For popularizing animal husbandry activities, promotion of animal husbandry programes are highly essential. The members of WDCS expressed their dissatisfaction on organization of extension programmes in the area either by the department or by the WDCS. The way they like the extension programmes to be conducted are documented are as follows.

After analyzing the data it is found that 78.18 per cent of the respondents expressed that extension programmes should be developed by the members of the society followed by 75.45 per cent preferred demonstration method over other methods. 80% of the respondents opined for the appointment of a dairy extension officer in the society.

7. Improvement of efficiency in functioning of WDCS

The WDCS members were interviewed in order to know about their opinions as to how the efficiency in functioning of WDCS could be improved. They expressed that functioning of society needs further to be transparent and there should be some conflict management strategy to handle the day to day problems of the society. Their suggestions were documented and as below.

The data reveal that of the total members 90% of the members opined to have honest executive members followed by transparency in society activities (83%), and society should have a conflict management strategy (82%).

XI. Overall Suggestions Given by WDCS Members

An attempt is made to compile all the suggestions given by the members of WDCS in order to know their preferences for improvement of functioning of WDCS. The findings are documented and analysed in Table 5.21.

Table. 5.21 : Overall suggestions given by WDCS members.

Suggestions	Average(%)	Rank
Improvement of availability of green fodder	80.90	I
Improvement of efficiency in function of WDCS	79.99	II
Improvement of milk procurement in society	75.81	III
Promotion of extension programme	75.45	IV
Improvement of availability of cattle feed	73.33	V
Improvement of availability of veterinary services	71.81	VI
Preparation and sale of milk product	54.54	VII

It is revealed from the above table that the members gave top most priority (80.90 %) for the improvement of availability of green fodder followed by bringing an improvement in function of WDCS (79.99 %), improving milk procurement (75.81 %). Increasing promotion of extension programme (75.45 %), making availability of cattle feed (73.33 %), making availability of veterinary services (71.81 %) preparation and sale of milk product (54.54 %) in order.

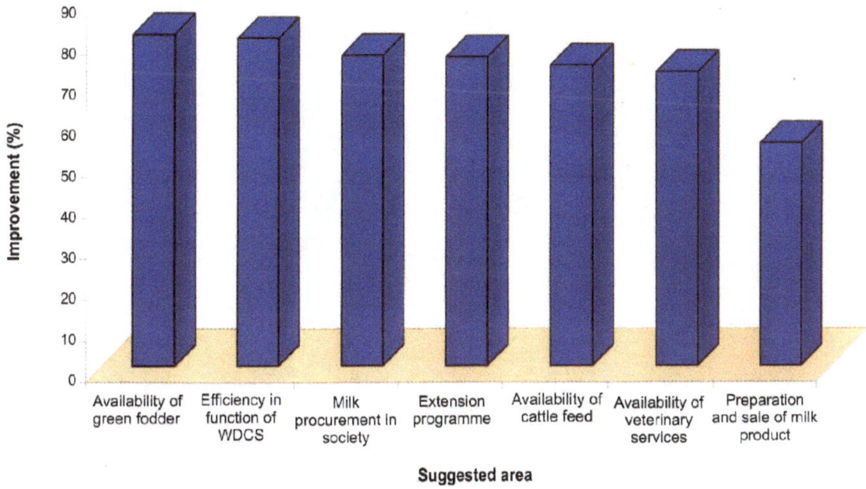

Fig. 5.22 : Overall suggestions given by WDCS members for improvement of society

❑❑❑

Bibliography

Agarwal, K. (1987). A study on contribution of farmwomen in animal husbandry enterprise. M.Sc. thesis (Submitted to) IVRI, Izatnagar.

Annual report (2003-04). Department of Agricultural Research and Education, ICAR, New Delhi.

Anonymous (2003). Indian Cooperative Movement at a glance. Swadeshi Patrika,New Delhi.

Arunachalam, J. (1999). Role of women in cooperatives. Perspective for 2000AD, in VMNICM (ed), Positioning Cooperatives in 21st century, VMNICM, Pune.

Balishter, Singh, R.P. and Singh, R. (2001). Rural development and corporatives. *Yojana*, **45**: 16-22.

Baviskar, B.S. (1988). Dairy Cooperatives and Rural Development in Gujarat. In D.W. Attwood and B.S. Baviskar (eds.) Who Shares Cooperatives and Rural Development 345-361, New Delhi. Oxford University Press.

Bhatt, E. (1975). Social Networks, *Human Relation* **8**(4):345-384.

Bhogal, T.S., Sharma, J.S. and Arora, V.P.S. (1988). Augmenting income and employment of small and marginal farmers and landless agricultural labourers through dairying in western Uttar Pradesh. *Indian Journal of Dairy Science.* **41**(1): 101.

Bhoite, H.S. and Shinde, S.B. (1987). A constraint analysis in relation to different components of dairy and management practices. Indian *Journal of Animal Production and Management*, **3**:166-177.

Bhuiyan, A.K.M.(1988) women's participation in agricultural tasks. An immerging trend in the changing traditional society of Bangladesh. . Paper presented in International conference on appropriate agricultural technologies for farmwomen,ICAR, New Delhi, 30th Nov. to 4th Dec.

Birader, R.D. (1986). Changing through dairy development - a case study. Dairy guide, 8(3): 2-24.

Chinnadurai, S., Chinnadsurai, P. and Markanday, J.C. (2002). Farm women in commercial dairy farming. Journal of Dairying *Foods and Home Science*, **21** (1): 63-65.

Chinnadurai, S., Chinnadurai,P., and K. Singh (2004). Relationship between socio economic characteristics and knowledge level of women prating Dairy farming. Indian Res. *Journal of Ext. Education.* **4**(1&2).

Dak, I.M., Sharma, H.L. and Jain, R. (1986). Social and Institutional framework of female participating in agriculture. *Indian Journal of Social work,* **47**(3): 286-287.

Dash, B. (2000). Role of women in agriculture. *Yojana,* **44**(11):35-37.

Desai, D.K. and Patel, K.V. (1974).Dairying as instrument of economic change. A case study of the leading modern dairies in India. 19th International Dairy Congress, New Delhi 64-75.

Dey, A.K. (1977). A study of decision-making pattern in rural families with respect to selected dairy husbandry practices in Chakka block of Allahabad district of UP. M.Sc. Thesis (Submitted to) NDRI, Karnal.

Dixit, V.B. and Laharia, S.N. (1995). Existing and desired level of interaction among dairy farmers. *Indian Journal of Extension Education,* **31**(1&4) : 30-45.

Dubey, V.K., singh, S.B.and Khera, J.K. (1982). Role of rural women in decision making with respect to animal husbandry practices. *Indian Journal of Home science.* **14**(2): 18-21.

Franco, F. and Shery Chand, P.G. Viajaya (1991).Operation Flood and The Voluntary Sector: A study of Tribal Dairy Cooperatives in Southern Gujarat. Indian Social Institute, New Delhi.

Ghosh, S. and Chand, R. (2001). Constraints in adoptin of recommended technologies for improved dairy farming. *Indian Journal of Extension Education,* **37**(1 & 2): 24-29.

Gosain, D.K., Sindhu, D.S. and Singh, R.R.B. (2002). Feed back analysis on the tested dairy processing technologies in Haryana. *Journal Dairying, Foods and Home Science,* **21**(1):52-55.

Gupta, S.N., Singh, R.I. and Shukla, S. (1986). Role of women in cattle care. *Dairy Guide,* **8**(11): 43-46.

Gurumoorty, T.R.(1995). Roles of cooperatives in Rural Development, *Kurukshetra* , November, 1995: 9-10.

Hirenkanagoudar, L.V. Hanumanthappa, H.S. and Jalihal, K.A. (1988). Impact of dairy development on the weaker sections. *Kurukshetra,* **36**(5):7-11.

Jain, V. and Verma, S.K. (1992). Nature and extent of involvement of men and women in animal husbandry operations. *Indian Dairyman,* **44**(7): 332-337.

Jamal, S. (1989). A study of role dynamics of farmwomen in relation to productivity of dairy animals. Ph.d. Thesis submitted to IVIR, Izatnagar.

Jhasi Rani, G. (1981). Decision making pattern of among rural families and type of participation by rural women in selected areas of decision. M.Sc. thesis submitted to APAU, Hyderabad.

Joshi, A.P. and Rawat, K. (2001). Choice of technologies for women. *Kurukshetra,* **50**(1): 20-24.

Khedre, R.L. and Pal, M. (1992). Role of farm women in dairy development. *Indian Dairyman,* **44**(7):12-24.

Kumar, A. and Sidhu, D.S. (2002). Impact of Tirhut Milk Producers'. Cooperative Union Limited on the knowledge of dairy farmers of Bihar. Journal of Dairying, *Food and Home Science,* **21**(2):128-131.

Kumar, G.S., Pandian,A.S.S. and Meganathan, N. (2003). Women in livestock farming. *Livestock International,* **7**(1):5-6.

Kumar, K. and Shnehalata, M. (1974). Training needs of farm women in Haryana. Indian Journal of *Adult Education*, **35**(9): 72-76.

Kumar, R. Singh, J.P., Jha, S.K. and Kumar, A. (1999). Beneficiaries knowledge level and attitude of a propose village dairy cooperative and its activities, Rural India, May, 99.

Kumari, A., Verma, I., Sinha, I. and Srinivasan, V. (1988). The chunking point - A case study of farm women in dairying in Bihar. Paper presented in International conference on appropriate agricultural technologies for farmwomen, ICAR, New Delhi, 30th Nov., to 4th Dec.

Lale, Sarachchandnra and Rao, R. Jagannath (1996). Whose cooperatives and whose produce? The case of LAMPS in Karnatak. A research paper: 53-91. Institute for social and economic change, Nagarabhavi, Bangalore.

Laxmi, R.D., Nataraju, M.S. and Murthy, N. (1999). Women in animal production: An ex-post-facto analysis. *Agricultural Extension Review*, **11**(3):3-8.

Mahajan, N. (1997). Measurement of adoption behaviour. *Indian Journal of Extension Education*, **33**(1&2):7-23.

Mahipal and G.C. Rai (1978).Role Performance of the Leaders of Efficient and Non-efficient Multipurpose Cooperative Societies. *Indian Journal of Extension Education*, **14** (1&2): 43-47.

Mahipal and G.C. Rai (1980). Relationships of Role Performance with Selected Socio Psychological Characteristics of the Leaders of Multipurpose Cooperative Societies. *Indian Journal of Extension, Education*, **16**(3 & 4): 46-50.

Malathi, D. (1979). A study of the knowledge and participation of rural women in agricultural operations with respect to paddy crops and their value orientation in Dharwad district. M.Sc. thesis. University of Agricultural Sciences, Dharwad campus, Karnataka.

Mankakar, Kamala (1990). Women and Employment . *Kurukshetra*, **52**(9): 30.

Mishra, R.V. (1999). Cooperatives for empowerment of people. *The cooperative*, **19**(12): 195-197.

Mohapatra, B.K., Mohapatra, P.S. and Nayak, M.P. (1996). Attitude and knowledge level of dairy farmers on practices of A.I. *Journal of Extension Education*, **1**(1&2): 101-103.

Nainta, R.P. (1995). Administration of Cooperatives in Himanchal Pradesh, *Kurukshetra*, November, 1995.

Nanavaty, R. (2000). Women's leadership in dairy sector. *Kurukshetra*, **49**(1):65-69.

Nataraju, M.S. and Channegowds, M.B. (1984). Differential adoption of improved dairy practices by small farmers, marginal farmers and agricultural labourers. *Indian Journal of Agricultural Extension Education*, **20**(3&4):63-66.

Nayak, H. S., S.S. Bajaj, and D.R. Rout (1986). Problems of Dairy farmers in Ahmadnagar District of Maharastra. *Maharastra Journal of Extension Education*. **5**:107-111.

Ogunbameru, B.O. (1990). Livestock extension service in the arid regionof Nigeria: A cast study of Borneo state. *Indian Journal of Extension Education*, **26**:17-23.

Om Prakash (1988). A study of training needs of farm women in scientific dairy farming practices in Karnal district of Haryana, M.Sc. thesis (unpublished), Kurukshetra University, Kurukshetra.

Padmanabhan, B.S. (2001). Women empowerment in farm sector. *Yajana*, **45**(1): 47-50.

Pandian, P. and Eswarn, R. (2002). Empowerment of women through micro-credit. *Yojana*, **46**:47-50.

Patel, B.T.(1980). A study of factors influencing successful functioning of Village Milk Producers' Cooperative Societies in Kaira District of Gujarat. An unpublished Ph.D. Thesis, IARI, New Delhi.

Patel, S.M. and Pandey, M.K. (1976). Economic impacts of Kaira district cooperative milk producers union (Amul Dairy) in rural areas of Kaira district (Gujarat State). Ahmedabad. Institute of Cooperative Management, p.10-19.

Puri, S. (1971). Rural families and decision-making pattern. *Indian Journal of Extension Education*. **7**(1&2): 65-67.

Rabishankar, A. and Thimmaiah, A. (1995). Cooperatives and Liberalization Challenges and Opportunities, *Kurukshetra*, November, 1995.

Rae, A.N., W.A., Wilson, W.R. Schroder (1985). Dairy Development, Trade and Policy issues within the pacific Basin. Agricultural Policy Paper No.11. Department of Agricultural Economics and Farm Management. Massey University. In: *Dairy Science Abstract*, , 48(1-6): No.244, 28.

Rangenekar, S.D.; Vasani, P. and Rangnekar, D.V. (1991). Initial report of study on women in dairy production. Proceedings of International workshop on "Feeding of ruminants of fibrous crop residues" of Indi-Dutch bioconversions project held at National Dairy Research Institute, Karnal.

Rangnekar, D.V. (1993). Traditional livestock production practices based on Indigenous technologies presented in National Seminar on "Indigenous Technologies for suits Annabel Agricultural Development at ICAR.

Rao, R. (1976). A study of the influences of selected factors and rural women's participation in agricultural tasks. M.Sc. thesis. Lady Irwin College, University of Delhi, New Delhi.

Rao, V.M. (2003). Women dairy cooperatives in Rajasthan, success story of a women dairy project. *Kurukshetra*, **51**(6): 19-22.

Rao, V.M. (2004). A worm's eye view of rural credit scenario in India in NAFSCOB (ed.) souvenir on 100 years of cooperative movement , 1904-2004, NAFSCOB, Mumbai.

Rao, V.M. (2004). Entrepreneurship through cooperatives. An experience of women's Dairy Projects in India. *Kurukshetra*, October, 2004, 52 (12):50-54.

Rao, V.M. (2004a). Empowering farm women through dairy cooperatives: A study in Andhra Pradesh and Karnataka . Anmol Publications, New Delhi.

Rao, V.M. (2004b). Changing destiny of rural women through dairy cooperatives in Bihar. Mittal Publication, New Delhi.

Rao, V.M. (2004c) Capacity building through women dairy cooperatives in Orissa. Reliance Publications, New Delhi.

Rao, V.M. (1981). Impact of dairy enterprise on income and employment potentialities. Dairy Guide 5:23.

Reddy et al. (1994). Women and rural development . A study of DWCRA in Chuddapah District, Kurukshetra, vol.XLII, No.12.

Sagar, R.L., Kunzru, O.N. and Singh, B. (1989). Productivity of dairy cows as related to management attributes of milk producers. Proceedings of 76th session of Indian Science Congress, Madurai.

Sah, A.K. and chand, R. (1999). Adoption of improved dairy practices in Bihar. *Indian Journal of Extension Education*, **35**(3 & 4):235-244.

Sah, U., Kumar, S., and Kumar, R. (1999). Impact of scientific training on farmwomen of KVK. *Indian J. Dairy and Bio Sc.* **10**(1&2).

Sambrani, Shreekant (1982). Managing commodity system: The development of cooperative Dairying in India. Agricultural Administration.

Sangwan, V; Grover, I. and Munjal, S. (1986). Participation of rural women of Haryana in Home. Farm and Dairy sectors. Paper presented in seminar on "Role of Developmental programmes on Socio-economic status of women held at Haryana Agricultural University, Hissar from March 21-22.

Sasikumar, M.V. (1988). Dairy Cooperatives in the Past Operation Flood Phase. *Yojana*, November, 1998: 31-33.

Sedhian, G.K. and Singh, B. (2001). Key performance area of village milk producer's societies in the hill regions. *Indian Journal of Extension Education*, **37** (1&2): 15-23.

Sen, D. and Rani, J.G. (1990). Women in dairying: A case study. *J. Rural Dev.* **9**(5): 809-831.

Sharma, K. and Saini, G.S. (2002). Reactions of the dairy farmers regarding the services provided by the milk cooperative societies in Ludhiana district of Punjab, *Journal of Dairying. Foods and Home Science*, **21**(1):18-22.

Shukla, R.K. and Brahmankar, S.D.(1996). Impact evaluation of operation flood on rural dairy sector. National Council of Applied Economic Research, New Delhi.

Singh, A.K. and Tyagi, K.C.(2001). Performance of milk cooperative in Haryana and Gujarat. *Indian Journal of Extension Education*, **37**(1&2): 7-14.

Singh, B.K.; Malhi Pal and K.C. Tyagi (1985). Study on Selected Socio-Personal-Economic Characteristics of Progressive and Non Progressive Villages of Karnal, District. *Indian Journal of Agricultural Extension Education*, **21**(3&4): 100-103.

Singh, L.P. (2004). Cooperative marketing in India: Status, constraints and correctiveness. A topic published in Kurukshetra – *A Journal of Rural Development*. **52**(12):39.

Sinha, M.N. (1971).Differential Adoption of Dairy Production Innovations. An Unpublished Ph.D. Thesis. Department of Extension Education, PAU, Ludhiana

Sohi, J.S. and R.L. Kherde (1980): A study of Dairy Adoption Behaviors of Small and Marginal Farmers in Pubjab. *Indian Journal of Extension Education*, **16**(1 & 2): 84-86.

Somjee, A.H. and Somjee, G. (1974). Cooperative dairying and the profiles of social change in India. Paper presented at 19th International Dairy Congress, India.

Sudharani, K., Sreelatha, K.K. and Surenda, G. (2000). Empowerment of women in rural areas. Rural India, October, 2000.

Sudharani, P., Raju, V.T. (1991). Participation of women in agril operation. *Indian J. Ext. Edu*. **27**. 54-59.

Thakar, R.F. and K.F. Patel (1988). Knowledge of farm women about improved agriculture under different land holdings, *Indian Journal of Social Research*, **32**(3): 272-276.

Thakur, D.S. (1975). Impact of dairy development through milk cooperative. A case study of Gujarat. *Indian Journal of Agricultural Economics*, **30**(3): 83.

Thiagarajan, M. (1984). The status of dairy farming in India, Dairy guide, 6:29-33.

Tiwari, M.K. (2001). Profitable milk business through dairy cooperative. A case study. *Agricultural Extension Review*, **13**(2):6-8.

Tripathy, H. (2001) Information receiving behaviour of rural women influencing productivity of milch animals. *Agricultural Extension Review*, **13**(4):12-17.

Tripathy, H. and Kunzru, O.N. (1992). Difference in socio-personal and socio-psychological characteristics of rural women between member and non-member dairy cooperative systems. *Indian Dairyman*, **44**(10):485-488.

Tyagi, K.C. (1975). Factors influencing the adoption of dairy innovations by farmers of ICDP. Karnal. Paper presented at Summer Institute, NDRI on modernization of dairy farming.

Usha, R. and Singh, C.B. (1982). Economic performance of farm women of weaker section in dairy enterprise. *Asian Journal of Dairy Research*, **1**(3&4): 206-212.

Veena, S.I., and S. Kunjal (1986). Participation of rural women of Haryana in home, farm and dairy sector: A report on ORP, NDRI, Karnal.

Venkateswaran, S. (1992). Living on the Edge: Women, Environment and Development, Friedrich Ebert Stiftung, New Delhi.

Verma, R.C. and Pant, D.C. (1978). Potentialities of increasing farm income and employment through dairying. *Indian Journal of Agricultural Economics*. **33**(3): 83-93.

Verma, T., Verma, S.K. and John, V. 91993). Training impact of improved practices on rural women. *Indian Journal of Extension Education*, **19**(1&2): 116-119.

Vinay, M.R. and Manjapa, D.H. (2004). Dairy cooperatives: Vital role in rural economy published in Kurukshetra. *A Journal on rural development* **52**(12): 55-60.

Vishwanathan (1989(. Women in livestock in India. A desk study of FAO Regional office for Asia and Pacific, Bangkok, Thailand.

Vyas, H.V. and Patel, K.F. (2000). Constraints faced by milk producers in adoption of dairy technology. *Agricultural Extension Review*, **12**(1):23-25.

Yadav, K.D. and Jain, N.C. (1984). Study on socio-economic factors influencing the adoption and non-adoption of crossbred cattle in western M.P. *Indian Journal of Extension Education*, **20**(1&2): 71-72.

❑❑❑

Interview Schedule for Respondents

Sl. No. ________

Dist. Milk Union .... Milk Route ....

Name of WDCS .... Village ....

Interviewed by : .... Interview date : ....

A. Socio-personal characteristics of women dairy farmers

1. Name of the Respondent : ____________________

2. Age in completed years : ____________________

3. Respondents' Education :

1. Illiterate
2. Primary
3. Middle
4. High School
5. Intermediate
6. Graduate
7. Post Graduate

4. Spouse's Education

1. Illiterate
2. Primary
3. Middle
4. High School
5. Intermediate
6. Graduate
7. Post Graduate

5. Family type :

(1) Nuclear (2) Joint

6. **Family size :**

No.	Relationship with respondent	Age (year)	Education	Main occupation

7. **Caste/ Tribe :** (1) SC (2) ST (3) OBC (4) General
8. **Religion :** (1) Hindu (2) Muslim (3) Buddhist
   (4) Christian (5) Jain (6) Other, specify
9. **Main Occupation of the family:**
   (1) Wage earning (2) Caste occupation
   (3) Business (4) Service
   (5) Farming (6) Dairy
10. **Total operational land holding (acres) :**
    a) Who owns land : Self / Husband / Others

Land use under	Area	Irrigated	Non-irrigated
Crop production			
Fodder production			
Uncultivable			
Total			

11. **Herd size**

Number of animals owned by age, class and breed of animal				
Description	**Desi**	**Crossbreed**	**Buffalo**	**Total**
Milch animal				
Dry animal				
Bullock/oxen				
Heifer				
Calves				
Total				

12. Present milk production/consumption/sale (litres)

Sl. No.	Types of milch animal	Per capita Milk production/ day	Total milk production/ day	Household consumption/ day	Milk sold to WDCS/day	Milk sold outside/ day	Unsold milk/day
1	Desi cow						
2	Cross breed						
3	Buffalo						
	Total						

13. Type of house you own :(1) Pucca (2) Semi pucca (3) Kutcha

14. Type of animal shed:

Structures	Within residential complex	Outside residential complex	Space		Ventilation		Location		Cleanliness	
			Adq.	Not Adq.	Adq.	Not Adq.	Suit.	Not suit.	Satis.	Not Satis.
Pucca										
Semi pucca										
Kuccha										

15. Farm powers : What do you have ?

(1) Bullock (2) Buffalo (3) Tractor

(4) Power tiller (5) Electric operated powers (6) Others

B. Procedures, formation and functioning of WDCS as per bye-law

16. Objectives

Each WDCS is a registered one. It has specific objectives. To what extent they are followed in your society ?

Sl. No.	Statements	Completely followed	Partially followed	Not followed
1	Attempt to produce quality milk under guidance of society			
2	Attempt to make milk marketing as profitable			
3	Improving breeds and health of animals.			
4	Sale of nutritionally balanced cattle feed.			
5	Fodder cultivation			

16.1 Funds

Society needs funds. What steps are taken for raising funds ?

Sl.No.	Statements	Followed	Not followed
1	Issue of shares		
2	Membership fee		
3	Receiving deposits		
4	Raising of loans		
5	Donations		

16.2 Membership

Membership of WDCS is essential and important. What criteria are followed to select members for the society.

Sl.No.	Statements	Followed	Not followed
1	Residence within operational area		
2	Making application in prescribed form		
3	Having buffalos/cows for milk production		
4	Supply milk to WDCS for three consecutive months		
5	Payment of membership fee		
6	Written agreement for supply of milk to the society only.		
7.	Not declared as bankrupt		
8.	Not convicted for crime		

16.3 General meetings

Holding of general meetings and other allied functions are the regular feature of the society. How far these are regular in your society?

Sl.No.	Statements	Followed	Not followed	Do not know
1	Annual General Meeting is held within period of three months after the end of each cooperative year.			
2	General meeting elected Chairperson .			
3	Declares members of the management committee.			
4	Sanctions budget for the next year			
5	Makes necessary arrangements for collection of milk			
6	Records proceedings in minute book			
7	Hold special General Meeting, on receipt of requisition from members			

16.4 Managing committee

(1) WDCS is expected to be regulated by general meeting, special meeting and management committee. In the opinion how regular these meetings are in your society.

(i) Very much regular (ii) Regular

(iii) Not regular

(2) Does your society has managing committee : Yes/No

(3) Are the members of managing committee decided as per bye-law (Yes/No)

16.5 Qualification of members of managing committee

Does your society follow criteria in deciding qualification of members of managing committee ? If 'yes', which of the followings are followed ?

Sl.No.	Statements	Followed	Not followed
1	Checking eligibility for membership		
2	Check whether defaulter on dues to society.		
3	Does not bears direct or indirect interest.		
4	Not disqualified by legal proceedings		
5	Not found liable to any legal offence of the society.		
6.	Supply milk regularly to the society		
7.	Any of the family members does not carry business to the society		

16.6 Duties and functions of the Chairperson

The chairperson of your WDCS has specific duties and function. To what extent you are satisfied in her performance.

Sl.No.	Statements	Fully satisfactory	Satisfactory	Not at all
1	Execution of resolution passed by the managing committee.			
2	Supervise routine work with secretary			
3	Deposit money in bank and draws money as per need in time.			
4	Verifies stocks, equipments and dead stocks of the society			
5	Regularly supervise works of society			
6.	Take action to rectify objections raised by audit.			
7.	Furnishes the detail information to additional register of the society.			
8.	Receives complains and disposes in time in judicious manner.			

16.7 Duties and functions of the Secretary

Secretary of the society is an important person for smooth functioning of the society. To what extent the Secretary of your WDCS performs duties and responsibilities.

Sl.No.	Statements	Fully satisfactory	Satisfactory	Not at all
1	Arranging general meeting, Managing committee meeting and other meetings when required.			
2	Spent, withdraw and recovers money as per instruction of Chairperson.			
3	Receives all communications and disposes the letters.			
4	Prepares all receipts, vouchers and reports.			
5	Maintain routine administration of society			
6.	Places audit report before managing committee in time.			
7.	Maintains Cash Book and other records of society.			
8.	Collects money and deposits in bank			

16.8 Distribution of profit

The society when gains profit it has to be distributed among the members as per respective shares. What is your opinion about distribution of profit in your society?

(i) Well managed (ii) Managed to some extent

(iii) Mismanaged

16.9 Election

Election to the society is a regular feature. The bye-law has to be followed for smooth managing of the election. In your opinion how the election procedures are followed on the following matters.

Sl.No.	Statements	Followed as per rule	Not followed
1	Notification of electoral role		
2	Notice of meeting		
3	Declaration of Vacancy		
4	Filling of nomination forms		
5	Voting		
6.	Counting of votes		
7.	Declaration of results		

C. Socio-economic impact on the members

17. Please state what changes are there on the items mentioned below and also specify the numbers/quantity whenever required.

Sl.No.	Content	Pre-WDCS	Post-WDCS
1.	Possession of Buffaloes		
	Possession of Improved buffaloes		
2.	Possession of Local cows		
	Possession of Crossbreds		
3.	Total production/day (litres)		
4.	Total consumption/day (litres)		
5.	Rate of milk (Rs./litre)		
6.	Requirement of extra manpower		
7	Materials possessed in numbers		
	i) Bicycle/bike etc.		
	ii) Tractor		
	iii) Pump set		
	iv) TV/VCR/VCP etc		
	v) Area covered for fodder production		
	vi) Others, specify		

18. On what aspects you have realized changes after being the members of WDCS ?

Sl.No.	Statements	Yes	No
1.	Annual income increased		
2.	Land purchased		
3.	Health care improved		
4.	House purchased or built		
5.	Children education improved		
6.	More accesses to mass media		
7.	More contact with AH extension functionary		

19. How do you invest your extra income earned from dairying?

Sl.No.	Statements	Yes	No
1.	Spend in the business		
2	Make saving		
3.	Spend for purchase of land		
4.	Spent for purchase/construction of house		
5.	For better health care		
6.	For social functions like marriage of sons/daughters		
7.	Mass media		
8.	Spend for education of children		
9.	Extension contacts		

20. Have you participated in any one or more extension activities - Yes or No. If yes, then state the following.

Sl. No.	Type of extension programme attended	No. of times attended/ last two years
1	Demonstration	
2	Training	
3	Exhibition	
4	Krishi Mela	
5	Field days	
6	Village meeting	

D. Managemental practices of dairy animals by members of WDCS

21. Have you grown any improved fodder crops in your field ? - Yes/No

Sl. No.	Types of fodder crops grown	Yield/ acre	Growing season	Sources of seeds/ saplings
1				
2				
3				
4				
5				
6				

22. What is the general practice of breeding animal in your area ?

Sl. No.	Practice of breeding	Always	Sometimes	Never
1	Mating with improved bull			
2	A.I.			
3	Let loose for mating with stray bulls			

23. How much do you pay for the service of bulls or A.I. for the animals?

Sl. No.	Types of service	Cost of service (Rs./Service)	Frequency of service		
			Regularly	Occasionally	Never
1.	Natural service				
	a. Desi bull				
	b. Improved bull				
	c. Buffalo bull				
2	Artificial insemination				
	a. Cow				
	b. Buffalo				

24. What type of feeds you generally give to your animals?

Sl. No.	Types of feed	Frequency of Offering		
		Regularly	Occasionally	Never
1	Concentrate only			
2	Green fodder only			
3	Dry roughage only			
4	Concentrate + green			
5.	Concentrate + green + roughage			

25. How do you manage to get feed for animals ?

Sl. No.	Type of feeds	Sources		
		WDCS	Local trader	Any other
1	Concentrate			
2	Green fodder			
3	Roughage			

26. Sources of vaccines and frequency of vaccination

Sl. No.	Sources of vac. Purchase	Types of vaccine available	Frequency of vaccination		
			Regularly	Occasionally	Never
1	Government vet. Hospital				
2	Cooperative store				
3	Private vet. Medical dealer				
4	Private vet. Practitioner				
5	Village live stockmen				
6	Others				
7	Not available in village				
8	Vaccinated against diseases				

27. How much you pay for the services of livestock technician/ veterinarians in village ?

Sl. No.	Types of services	Fee for each visit	
		Veterinary doctors	Technician
1	Minor first aid/visit		
2	Major treatment		
3	Vaccination		
4	Dystokia		
5	Drenching/dusting		
6	Others		

28. Who does the following works of dairy management in your home?

Sl. No.	Types of services	Who					
		Self	Husband	Servant	Self + Hus	Self + Servant	Husband+ servant
1	Arranging green fodder						
2	Arranging concentrate						
3	Feeding of animals						
4	Watering animals						
5	Milking animals						
6	Taking milk to cooperative						
7	Washing the milking trough						
8	Collecting the dung from shed						
9	Grooming the animal						
10	Taking the buffalo for wallowing						
11	Taking the animal for mating/ A.I						
12	Taking the animal for treatment						
13	Who decides to sale of animal						
14	Who decides purchase of animals						
15	Collection of money from milk sale						
16	Purchase of medicine etc.						

E. Key performance areas of WDCS

Perceived performance of key activities of WDCS. The respondents are required to carefully read/listen the following statement/ questions and give their judgment in yes/no forms.

Sl. No.	Key performance areas in WDCS	Yes	No
1.	Procurement of milk is regular.		
2.	Members are satisfied in milk sale		
3.	Milk procurement is in increasing trend		
4.	Milk procurement is non-break		
5.	Payment is regular to pourer		
6.	Payment of union is regular to WDCS		
7.	Union is satisfied on quality and price of milk.		
8.	WDCS sales milk to local consumers		
9.	Milk is also sold to local consumers by members		
10.	Milk products are also sold by WDCS		
11.	Milk products are remunerative		

Sl. No.	Key performance areas in WDCS	Yes	No
12.	WDCS purchase cattle feed from union and sales to members		
13.	WDCS sometimes procure cattle feed from outside.		
14.	Emergency service is needed sometimes		
15.	Veterinary doctor is always available.		
16.	Members are satisfied with service of Veterinary Doctor in terms of treatment and fee.		
17.	WDCS fixes fees of Veterinary Doctor		
18.	Cows are regularly vaccinated.		
19.	All animals of the members are vaccinated in time.		
20.	WDCS purchases vaccines from union		
21.	Sometimes WDCS purchases vaccines from outside.		
22.	WDCS vaccinate the animals at free of cost		
23.	Members are satisfied on quality of vaccines.		
24.	There is easy access of loans to the farmers.		
25.	The interest rate of commercial/development bank is higher than private money lenders in the village.		
26.	It is not easy to get loan because needs lot of procedures and needs to satisfy the people involved in between.		
27.	WDCS is fully equipped with first-aid equipments		
28.	Members bring cow to WDCS for first-aid		
29.	Provision is there to render first aid at members door step.		
30.	Training programmes are conducted regularly by WDCS.		
31.	Training programmes are relevant and need based.		
32.	Members satisfied with training Programme.		
33.	Training programmes are free of cost to the members.		
34.	Members are satisfied on A.I. services of WDCS .		
35.	A.I. services are timely, regular, affordable and qualitative.		
36.	Members are interested for service of breeding bulls for their cows in heat.		
37.	Adequate number of breeding bulls are provided by WDCS		
38.	WDCS maintain the breeding bulls at society headquarters and members used to take the advantage		
39.	Green fodders are available in village		
40.	Seeds and planting materials are adequately available to the farmers by the WDCS.		
41.	Fodders are grown throughout the year		
42.	Members are also knowledgeable on fodder cultivation		
43.	Fodder yield per hectare of land is not satisfactory.		
44.	Dairy extension programmes are conducted regularly by WDCS		
45.	Farmers are fully satisfied on the type of extension service provided by WDCS		
46.	Extension programmes are need based		
47.	Farmers participation in the extension programme is encouraging		

F. Entrepreneurial characteristics of WDCS members

The followings are some of choices. State your remarks on suitable alternatives which you like the best.

1. For survival in the society one has to work hard. When such situation arises how do you face the situation ?
   a) Work more taking it as challenge
   b) Wait for others help
   c) Leave it and think for alternation
2. Every body thinks of achieving some thing in life. Have you ever thought of .
   a) To be rich
   b) To gain social status
   c) To be in high position
   d) Not yet thought in that line
3. When you are keeping her you may have some problems. When your cows suffer or do not give milk as expected what do you ?
   a) To stop of keeping cows
   b) Get disgusted and continue without interest
   c) Promptly take action to improve
4. When you started keeping cow for selling milk what was your ideas ?
   a) Just keep the cows
   b) Anticipated the problem
   c) Anticipated how to increase the profit
   d) Did not think anything
5. When you took up diary, how did you decide ?
   a) You decided yourself
   b) Sought advice of others to start
   c) Because others persuaded you.
6. Did you ever thought of about diary farming in terms of
   a) Market demand
   b) Consumer preference
   c) Did not think such matters.

7. Do the people of your village feel you as
   a) Very good organizer of WDCS
   b) Just a member
   c) Leader in going forward
8. How do you think yourself in terms
   a) First to try an idea for business
   b) Follow others in business
   c) Last to join in the society after being confirmed about its benefits.
9. There are many new ideas/technology in the business. What is your position in seeking new information ?
   a) First to enquire
   b) Contact different people
   c) Contact media
   d) Wait and listen from the villagers.
10. How far you are quality conscious in your milk supply to WDCS
    a) Very much
    b) Much
    c) Follow others

G. Problems faced by the members of WDCS

As member of WDCS and doing milk business what problems do you face ? Please tick the items you feel appropriate.

1. In arranging finance
2. Non-availability of good breed/feed/fodder
3. Competition high in market.
4. As women your mobility is limited.
5. Family household work is too much.
6. Lack of education.
7. Because you are a women.
8. You do not possess risk bearing capacity.
9. WDCS is not well functioning.
10. Union people are not helpful.
11. Any other problems not mentioned above.

H. Suggestions for improvement of WDCS

Please give your suggestions under each statement.

1. To improve procurement of milk from members

2. To prepare and sale milk products.

3. To make cattle feed available

4. To grow green fodders.

5. To avail efficient veterinary services

6. To promote extension programmes in dairy.

7. To run WDCS efficiently.

□□□

FSC
www.fsc.org
MIX
Papier | Fördert
gute Waldnutzung
FSC® C083411

Zeitfracht Medien GmbH
Ferdinand-Jühlke-Straße 7
99095 Erfurt, Deutschland
produktsicherheit@kolibri360.de